DETERRENCE BY DENIAL

Deterrence by Denial

Theory and Practice

EDITED BY

Alex Wilner and Andreas Wenger

Rapid Communications in Conflict and Security Series
General Editor: Geoffrey R.H. Burn

CAMBRIA
PRESS

Amherst, New York

Front cover image: A medium-range ballistic missile target is launched from the
Pacific Missile Range Facility in Kauai, Hawaii, during Flight Test Standard Missile-27,
Event 2. (U.S. Navy photo by Latonja Martin/Released)170829-O-WE597-723.

Library of Congress Cataloging-in-Publication Data

Names: Wilner, Alex S., 1979- editor. | Wenger, Andreas, editor.

Title: Deterrence by denial : theory and practice / edited by Alex S.

Wilner & Andreas Wenger.

Description: Amherst, New York : Cambria Press, [2021] |
Series: Rapid communications in conflict and security series |
Includes bibliographical references and index. |
Summary: "Deterrence by Denial: Theory and Practice is the first study to focus
exclusively on contemporary denial, bridging the theoretical gap that persists between
classical deterrence theory and contemporary insecurity. The book significantly
advances the scholarship on deterrence by denial with empirically driven and policy-
relevant contributions written by leading international scholars of conventional
military aggression, missile defense, terrorism and militancy, crime, and cybersecurity.
Deterrence by Denial: Theory and Practice is an important and unique book,
of interest to scholars of international relations, political science, terrorism and
intelligence studies, and cybersecurity, as well as to policy analysts, practitioners, and
members of the armed forces and intelligence community"-- Provided by publisher.

Identifiers: LCCN 2020036405 (print) | LCCN 2020036406 (ebook) |
ISBN 9781621965503 (library binding) |ISBN 9781621965510 (paperback) |
ISBN 9781621965909 (adobe pdf) | ISBN 9781621965916 (epub)

Subjects: LCSH: Deterrence (Strategy) | Access denial (Military science)

Classification: LCC U162.6 .D45 2020 (print) | LCC U162.6 (ebook) |

DDC 355.02--dc23

LC record available at https://lccn.loc.gov/2020036405

LC ebook record available at https://lccn.loc.gov/2020036406

TABLE OF CONTENTS

List of Figures

DETERRENCE BY DENIAL

DETERRENCE BY DENIAL

NEXT STEPS

Alex Wilner and Andreas Wenger

Within the study of international relations (IR), deterrence theory incorporates the core elements of cost and benefit into its theoretical foundation. Deterrence by punishment explores how threats of retaliation, military or otherwise, might dissuade, deter, compel, or altogether influence an adversary to behave in a particular manner. Manipulating behaviour does not solely revolve around the issuance of punitive threats, however. An adversary's actions are informed by the costs of pursuing that action *and* by the benefits the action is expected to provide. In the former case, costs can be augmented by threatening retaliation. The message is "If you decide to act in this particular way, we will surely respond in a manner that is painful to you." Unwanted behavior is curtailed because the promise of retaliation can at times be sufficient to diminish the value of an action by making it appear more costly. But in the latter case, costs are augmented by diminishing the expected gains the action is meant to provide. In this case, the message is: "You may want to achieve a certain goal, but we have developed ways to hinder, complicate, and diminish your progress and odds of success." Here an unwanted behavior is curtailed because the expected benefit the action is meant to provide an actor no longer appear certain; the contemplated action may not be worth it. In the nomenclature popular to IR and deterrence theory, this process is labelled deterrence by denial.[1]

In the theory and practice of deterrence during the Cold War, deterrence by punishment remained dominant. Yet at the level of Western grand strategy, the logic of denial had a far stronger influence than is commonly acknowledged. During the Cold War, the objective of policy (to uphold the bipolar status quo) and the choice of strategy (deterrence used to preserve the status quo through coercion) were well aligned: both reflected a logic of denial. The West's primary objective during the Cold War was the containment of Soviet-Communist influence around the world. Patrick Morgan asserts that the "Cold War world was seemingly deterrence dependent."[2] Western powers checked the perceived expansion of the Soviet-Communist bloc by building up and coordinating their defenses within a global alliance system. Thus, the principal aim of deterrence as the strategy of choice was to avoid nuclear war and to stabilize the strategic interaction between East and West. Deterrence was conceptualized as "the power to dissuade another party from doing something" and was differentiated from compellence, understood as the power to coerce another "*to do* something" it might not otherwise have done.[3]

How then do we explain the dearth of theoretical and practical attention to denial during the Cold War? The main reason why denial played such a limited role was that the Western powers, following classical deterrence thinking, saw nuclear punishment as the principal deterrent mechanism and, at the level of nuclear conflict, thus purposefully ruled out threats of denial. Denial was sidelined to lesser strategic questions of tactic and conventional warfare.[4] Given the threat of nuclear exchange during the Cold War, deterrence by punishment offered an abstract framework geared towards controlling the desired strategic outcome: The permanent elimination of large-scale military conflict between East and West. In such a symmetrical conflict between states armed with nuclear weapons, the goal had to be the absolute deterrence of any use of nuclear force, since the failure of deterrence would result in a nuclear holocaust. "The mere prospect of nuclear annihilation," Thomas Rid adds, "contaminated the West's conceptual, historical and operational understanding of deterrence."[5] Preventing nuclear war and escalation was paramount; deterrence became "synonymous with nuclear weapons."[6] As long as the aim was to deter nuclear attacks on the US homeland, US nuclear threats did not have to be very credible to be effective.[7] Far more complex, however, was the extension of the US nuclear deterrent over its European and Asian allies. Massive American nuclear threats to deter minor conventional probing by the Soviet Union in Europe lacked credibility. In case of a conventional attack by Warsaw Pact forces on NATO territory and on NATO forces, the key role of Western conventional forces was to ensure that the nuclear threat was credible. Extended deterrence was thought top-down and the primary goal of a conventional option in Europe was to reinforce NATO's position of escalation dominance.

Western nuclear deterrence theory during the Cold War was preoccupied with the problem of upholding stability, especially given the shifting military balance and during associated incidents such as the Berlin Crisis and the Cuban Missile Crisis. Strategic stability between the United States and the Soviet Union was anchored in mutual vulnerability, and this meant that there was no room for threats of denial at the strategic level. The key tools for upholding the status quo were threats of massive nuclear punishment. In nuclear deterrence, the credibility of capability was indisputable. What was questionable, however, was the mutual resolve to use nuclear weapons, given the unimaginable consequences of nuclear war. Accepting an offense-dominated military technological setting, the two antagonists mutually assured that they would retain their vulnerability to threats of nuclear punishment. Consequently, over time nuclear deterrence—reinforced by arms control—became a superpower tool for freezing the Cold War status quo, limiting the arms race, restricting the proliferation of nuclear weapons, and strengthening strategic stability.[8]

THE 21ST CENTURY: MANAGING COMPLEX SECURITY CHALLENGES

Things have begun to change; denial appears to be making a steady comeback. Deterrence by denial plays a growing role in the theory and practice of contemporary deterrence across the various domains of conflict. But it has taken a considerable amount of time for threats of denial to establish themselves firmly as one of the many tools available for managing an increasingly complex security environment. In the early post–Cold War years, the reduced influence of nuclear weapons and the apparent dominance of the one remaining superpower diminished the apparent need for deterrence by punishment; deterrence as a strategy came to be seen as somewhat redundant. Soon after, however, deterrence thinking—and, in particular, deterrence by denial—began to resurface, and with far greater intensity in US than European strategic circles.[9] This renewed interest in deterrence was driven by the transformation of the international security system and by emerging regional and global security challenges. As the international system became more fragmented, and as technological advances spread globally, Western policymakers and academics began to focus on two interrelated and increasingly relevant types of security contexts: On the one hand, the emergence of new regional and global powers, along with missile and WMD proliferation, and the associated repercussions for homeland security, regional (extended) deterrence architectures, and strategic stability among great powers; and on the other hand, the broadening and growing role of low conflict (insurgency, terrorism, cyberconflict) and the consequences for the management of global security affairs beyond the state.[10] All of this has created space for new ways of thinking about deterrence and coercion among and between states and non-state actors alike.[11]

The most important driver for the growing interest in the logic of denial in the context of more traditional strategic interactions among states and allies was the rapid diffusion of new technologies and the ensuing potential for long-range coercion. The diffusion of military technology, from a Western perspective, gave rise to three interrelated security challenges: First, the proliferation of ballistic missile and nuclear technology to regional powers such as North Korea and Iran created new potential long-range threats to homeland security. Second, and more importantly, the diffusion of ballistic military technology undermined the credibility of US security guarantees and the stability of regional deterrence regimes. And third, the development of anti-access/area-denial capabilities by China and Russia compounded the problem of US extended deterrence in regional settings. Given these three new contingencies, the West has shifted its deterrence thinking towards denial, with missile defense gradually becoming its most visible material component. US deterrence policies now exhibit a new triad of nuclear and non-nuclear offensive strike systems; active and passive defenses; and an integrated defense infrastructure. Although nuclear weapons continue to strengthen regional deterrence and reassure US allies, the role of conventional power-projection and effective theater ballistic defenses in US extended deterrence policies continues apace. NATO, too, in recent deterrence and defense posture reviews, integrates a denial component in the form of conventional military and new missile defense capabilities into the alliance's extended deterrence guarantees.[12]

Beyond interstate conflict and with respect to nontraditional and non-state threats and concerns, the growing interest in the logic of denial in the post–Cold War years was predominantly driven by the fact that, in general, threats of punishment only work when perpetrators are known or can be otherwise identified. In deterring terrorism and cyberattacks, such knowledge is often illusive; attribution (and retaliation) is often impossible. The process of repurposing deterrence theory to address these novel contemporary challenges has begun in earnest.[13] One central lesson that has emerged from this new scholarship is that deterrence by denial has become a more appropriate hedge against low and recurring conflict involving non-state actors, because it has the potential to limit the type and intensity of terrorism even if it cannot eliminate all threats of violence.[14]

In other important ways, the dimensions of war themselves have expanded: Cyberspace has joined land, air, sea, and outer space as an operational domain of conflict and warfare. Upon declaring NATO's intent to include cyberspace within its preview, NATO Secretary General Jens Stoltenberg reiterated in 2016 that it was "hard to imagine a conflict without a cyber dimension." As a result, General Stoltenberg added, NATO had "decided that a cyberattack can trigger Article 5" of the North Atlantic Treaty, which would initiate the alliance's collective defense mechanism.[15] Article 5 is the deterrent bedrock upon which the alliance has survived nearly seven decades. Theoretical inroads into cyber deterrence are now being sought.[16]

In the real world, strategic interactions between states, and strategic interactions between states and non-state actors, tend to overlap. This has given rise, in theory and practice, to new deterrence concepts that take into account the broader application of the strategy, as well as the need for a more tailored delivery. Recently, as part of the "fourth wave" of deterrence research, scholars have rediscovered and revisited the traditional tenets of deterrence theory in light of new and often asymmetric threats. In nontraditional asymmetric interactions between state and non-state actors, criminological deterrence theory has been used to reconceptualize deterrence as a dynamic process that involves multiple threats to multiple targets. Thus, in such settings, deterrence is increasingly seen as merely one tool among many military, diplomatic, and civilian tools that can be used to influence potential adversaries.[17] At the same time, with more traditional strategic interactions between states and allies, conventional deterrence theory has been rediscovered and reapplied to a situation in which non-nuclear precision-strike capabilities and multilayered missile defense systems are playing a growing role.

Given the wide range of regional multilateral security efforts, deterrence regimes are increasingly integrating a wide range of diplomatic and civilian tools.[18] In both settings, scholars and practitioners now emphasize the growing importance of denial strategies and often use these in combination with punishment approaches and/or broader political and economic influence strategies.

All told, contemporary security dynamics have changed dramatically since the end of the Cold War. Threats are increasingly sub- and non-state in nature, and they are much more diffuse. Whereas nuclear weapons and deterrence by punishment still matter in governing international crises and shaping interstate relations, non-state adversaries, conventional military challenges, digital-based threats, and threats short of open war have today cumulatively tipped the deterrent calculus in favor of denial. New conceptual ground long dormant during the Cold War is being uncovered, leading to new proposals of and discoveries in deterrence by denial. As a community of scholars and practitioners we find ourselves today at the rising dawn of denial theory and practice.

The logic of denial is simple: denial reduces the perceived benefits an action is expected to provide an adversary. Whereas punishment manipulates behavior by augmenting costs, denial works by stripping away benefits. By illustration, hardening defenses against attack—what Lawrence Freedman might dub "passive" defenses—raises the cost to a would-be challenger by diminishing the probability that they are likely to acquire their intended objective or goal.[19] So whereas punishment deters through fear of pain, denial deters through fear of failure. Beyond these basic tenets, however, contemporary deterrence by denial is not well understood. How might deterrence by denial be conceptualized and theorized? How does denial figure into contemporary military strategy and planning? How does denial relate to punishment in practice? How has the nature of warfare evolved to facilitate a resurgence in denial strategies? What technological innovations help empower denial over punishment? How can denial be used by both strong and weak states alike to acquire their objectives? How does denial work against, and on behalf of, non-state actors, including terrorist organizations, insurgents, and criminals? And what are the promises and pitfalls of practicing denial in cyberspace?

A ROADMAP TO THE VOLUME

The purpose of the volume is less about developing a singular theoretical or conceptual approach to denial that might, somehow, be applicable to a range of disparate threats, concerns and challenges stemming from states, non-state actors, and cyberspace. Rather, the volume is an attempt to free scholars to push and prod deterrence by denial—and the related concepts of influence, dissuasion, defense, and coercion—into the twenty-first century. At times the contributors build off each other, both in terms of developing theory and in building empirical evaluations. But at other times contributors disagree—even clash—suggesting that the logic of denial may come in different shades, more or less appropriate depending on circumstance and situation. Either way, our collective purpose is to nudge deterrence scholarship along, to provide new ideas on denial that crosses the traditional domains of warfare and security.

The chapters collected and presented here are the culmination of a lengthy research program. Our contributors first met during an author's workshop convened by Alex Wilner and Andreas Wenger (the editors of this volume), at the University of Toronto, Canada.[20] The Toronto gathering was the first of its kind to ask a disparate collection of security experts from a variety of fields to think about how the logic, theory, and practice of denial might apply to their specific areas of research and expertise. The volume, like the workshop itself, explores two overarching themes divided into two component parts: denial in theory and denial in practice.

In the case of denial in theory, Part One of this volume provides a theoretical assessment of contemporary denial, presenting insights and lessons derived from several fields and disciplines, including IR, strategic studies, terrorism studies, and criminology. Four chapters make up this section. In chapter 1, Patrick Morgan provides an overview of the conceptual building blocks of deterrence by denial, as it relates to developments in deterrence more broadly. Using an historical approach, Morgan illustrates the various ways in which denial has evolved over the past few decades to reflect the changing nature of interstate security and conflict. He provides a detailed summary of the logic of deterrence by denial, along with a thorough exploration of the advantages and disadvantages denial provides contemporary decision-makers. In chapter 2, Alex Wilner reaches back into traditional, Cold War–era deterrence-by-punishment theory to reconceptualize and repurpose certain coercive concepts for contemporary denial. He develops three unique concepts: intra-conflict denial, cumulative denial, and communicative denial. Using descriptive historical scenarios, Wilner illustrates how each concept might work in practice in different security contexts.

In chapter 3, Janice Gross Stein and Ron Levi provide a theoretical and empirical assessment of denial (and delegitimation) in contemporary counterterrorism, drawing on concepts and approaches developed by criminological deterrence theory. The authors provide conceptual lessons for building a cross-disciplinary approach to contemporary denial for tackling substate threats and challenges. Finally, in chapter 4 John Sawyer proposes a new definition of "preventative influence" in counterterrorism, which he terms *dissuasion by denial*. He then comparatively evaluates the dissuasive effects of defensive (or dividing) walls in Israel and Northern Ireland, drawing empirical conclusions from these two illustrative examples for practicing denial in counterterrorism more broadly. Sawyer then illustrates the methodological challenges associated with measuring coercion in these and other empirical cases.

In the case of denial in practice, Part Two of this volume offers an empirical assessment of contemporary denial, presenting lessons derived from interstate and regional conflict, missile defense, strategic culture, counterterrorism, and cybersecurity. Here, different models, theories, and approaches of denial are assessed in practice using relevant cases from around the world. Four chapters are included in this section. In chapter 5, James Wirtz offers an empirical assessment of US deterrence and denial, providing a typology of US defense strategies that dominate and characterize contemporary US military thinking and planning. Wirtz illustrates how and why contemporary adversaries of the United States have come to believe that they can defeat American deterrence, and suggests ways in which American decision-makers might reconceptualize US deterrence-by-denial strategies to augment the credibility of their overall deterrence posture. In chapter 6, Jonathan Trexel offers an in-depth exploration of Japan's evolving deterrent relationship with North Korea, providing a timely and unique, post–Cold War case study of a deterrent relationship that pits a nuclear-armed state against a non-nuclear (though highly advanced) rival. Using Japan's ballistic missile defense technology as a backdrop, Trexel illustrates the nature of Japanese denial and its effect on North Korean behavior, providing lessons for other countries and rivalries.

In chapter 7, Dmitry (Dima) Adamsky examines Israeli conceptualization of deterrence. He contrasts Israeli practice of deterrence with traditional or classical interpretation of IR deterrence common in the West, describes how Israeli deterrence archetypes evolved over time and as a direct result of regional crises and conflicts, and traces the evolution of denial strategy in Israeli strategic thinking and culture. He pays particular attention to Israeli efforts to combat, deter, and defeat militant organizations like Hamas. Finally, in chapter 8, Martin Libicki, explores the complexity of thinking about, applying, and practicing coercion in cyberspace. In terms of cyber denial, he illustrates that gauging success may depend in part on whether a defender is trying to defeat a cyberattack altogether, or simply to effect, limit, or dampen what a cyberattack is meant to accomplish. He applies this logic to high-end cyberattacks launched by one state against another in the realm of critical infrastructure (e.g., strategic cyberwar) and deployed forces in theatre (e.g., operational cyberwar).

In the conclusion, we, as editors of the book, provide a substantive and detailed review of the project's individual and comparative findings as they relate to the larger study of deterrence. Focus is placed along two lines: advancing the theory of denial within the sub-disciplines of IR, strategic studies, terrorism studies, and criminology; and assessing the effect of denial across various contemporary security domains to generate lessons for putting denial into practice. The purpose of our conclusion is to take stock of the field's recent advances, and to provide academics and practitioners alike with a guide for further exploring and applying deterrence by denial to contemporary security affairs.

FINAL THOUGHTS

For scholars of IR and for security practitioners, military planners, and policy analysts alike, updating deterrence by denial for contemporary threats and insecurities features all the elements of an exciting and generally understudied enterprise, pairing a classical literature to a set of evolving and complex security challenges. Whereas the fourth wave of deterrence scholarship has indeed begun to bridge the divide between classical deterrence theory and contemporary security concerns, specific issues like terrorism, organized crime, and cyber threats have nonetheless long rested outside the general purview of deterrence scholarship.[21] This book will advance our collective understanding of deterrence by denial in theory and practice by empirically applying the concept to both a variety of different contemporary threats, including conventional military aggression and missile defense, and to novel and emerging threats, like terrorism and militancy, crime, and cyber and digital security. In illustrating and assessing how theories of deterrence, both new and old, apply to contemporary challenges, our goal is to stake out new theoretical territory upon which the logic of denial might be expanded, and too advance new strategies and approaches for applying denial in practice.

NOTES

1. Snyder, *Deterrence and Defense*, 13–16.
2. Morgan, "The State of Deterrence in International Politics Today," 85–87. See also Patrick Morgan's contribution to this volume, chapter 1.
3. Snyder, "Deterrence and Power," 163; Schelling, *Arms and Influence*.
4. Snyder, *Deterrence and Defense*; Mearsheimer, *Conventional Deterrence*; Shimshoni, *Israel and Conventional Deterrence*.
5. Rid, "Deterrence Beyond the State," 125.
6. Gerson, "Conventional Deterrence in the Second Nuclear Age," 34
7. Schelling, *The Strategy of Conflict*, 6.
8. Wenger, *Living with Peril*.
9. See, for instance, Lanoszka and Hunzeker, "Confronting the Anti–Access/Area Denial and Precision Strike Challenge in the Baltic Region,"; Lanoszka and Hunzeker, *Conventional Deterrence and Landpower in Northeastern Europe*; Deni, "Modifying America's Forward Presence in Eastern Europe"; Beckley, "The Emerging Military Balance in East Asia."
10. Knopf, "The Fourth Wave in Deterrence Research"; Harknett, "The Logic of Conventional Deterrence and the End of the Cold War"; Wenger and Wilner, *Deterring Terrorism: Theory and*.
11. See, for instance, Morgan, *Deterrence Now*; Art and Cronin, *The United States and Coercive Diplomacy*; Freedman, *Deterrence*; Smith, *Deterring America*; Goldstein, *Deterrence and Security*; Paul, Morgan, and Wirtz, *Complex Deterrence*; and von Hlatky and Wenger, *The Future of Extended Deterrence*.
12. von Hlatky and Wenger, *The Future of Extended Deterrence*.
13. See, for instance, Davis and Jenkins, *Deterrence and Influence*; Shapiro, *Containment*; Lowther, *Deterrence*; Wenger and Wilner, *Deterring Terrorism*, and Wilner, *Deterring Rational Fanatics*; Payne, *Deterrence in the 2nd Nuclear Age*; Long and Wilner, "Deterring an 'Army Whose Men Love Death'"; Payne, *Strategy, Evolution, and War*; Lindsay and Gartzke, *Cross-Domain Deterrence*.
14. Wenger and Wilner, *Deterring Terrorism*.
15. North Atlantic Treaty Organization, Press Conference with Secretary General Jens Stoltenberg, June 14, 2016.
16. Libicki, *Conquest in Cyberspace*; Libicki, *Cyberdeterrence and Cyberwar*; Singer and Friedman *Cybersecurity and Cyberwar*; Mazanec and Thayer

Deterring Cyber Warfare; Slaughter, *The Chessboard and the Web*; Mandel, *Optimizing Cyberdeterrence*; Wilner, "US Cyber Deterrence."

17. Wenger and Wilner, "Deterring Terrorism: Moving Forward."
18. Morgan, "The State of Deterrence."
19. Freedman, *Deterrence*.
20. With thanks, we appreciate the kind support the workshop received from the Munk School of Global Affairs and Public Policy at the University of Toronto, Canada, the Center for Security Studies (CSS) at the ETH Zurich (Swiss Federal Institute of Technology), Switzerland, and Canada's Department of National Defence's Defence Engagement Program (DEP).
21. Lupovici, "The Emerging Fourth Wave of Deterrence Theory"; Knopf, "Terrorism and the Fourth Wave in Deterrence Research."

CHAPTER 1

DETERRENCE BY DENIAL FROM THE COLD WAR TO THE 21ST CENTURY

Patrick M. Morgan

After a period of modest neglect, deterrence has been reviving in importance and getting more attention. Initially it received serious study because of the emergence of the Cold War in hopes it would prevent another great war with cataclysmic results. As a result, eventually it was often cited as responsible for the survival of humanity. Missing from classic deterrence theory that developed was how deterrence would inevitably be shaped, in part, by the character of the international system and the domestic environments of the states involved. Instead, analysis gravitated toward depicting abstract environments and actors (e.g., states A and B, rational actors,) even as deterrence in fact also clearly reflected elements of the real environment: the intense Cold War, opponents' harsh conceptions of each other with intense hostility, seemingly poised to seize an opportunity to attack. Treating deterrence abstractly evolved as a deliberate contribution, in part, to tone down the conflict and ease the tensions.

When the Cold War temporarily dissolved into a suddenly altered international environment close to the end of the twentieth century, deterrence became much more recessed and of less interest to academic and other security analysts. But it soon returned to considerable prominence. Studying deterrence did so as well, because deterrence in practice has needed significant adjustments to reshape or refine deterrence theory to fit the fluid emerging international system that by now is on the verge of a new era that is beginning to once again rearrange how deterrence will be employed. Renewed attention to deterrence by denial, displayed in this volume, is a serious development along these lines. As Alex Wilner and Andreas Wenger note in their introduction to the volume, "denial appears to be making a steady comeback."[1] This book explores why and how.

Deterrence by denial involves threats, active and passive, designed to make a potential attack appear unlikely to succeed so as to convince the potential attacker to abandon it; plus the use of force to make a real attack unsuccessful causing the attacker to abandon it; and making a successful attack so difficult and costly (a pyrrhic victory) that no further attacks are mounted—making an attack unsuccessful via an effectively threatening defense or its effective implementation. Thus, deterrence by denial is more extensive than, but encompasses, deterrence by defense.

Why its importance today? For one thing it is more usable for facing the real, not just possible, threats which today are numerous but thankfully relatively limited in scale. And the threats are more often carried out, making them more readily studied, assessed, and hopefully destroyed, in contrast with Cold War deterrence tasks when it was often uncertain whether denial was really working or even needed and, if working, what aspect was most useful, most effective, under what conditions. Deterrence by denial of that era could be hard to assess with regard to controlled and tailored punishments effects, in a context of potentially immense harm. Was the other side really scared off, or just biding its time? (Escalation ladders were devised to cope with this problem but were abstract examples, not functional strategies enjoying wide confidence.)

Currently deterrence by denial is readily applicable when needed, being applied but failing all too commonly, for instance—for the United States—in Afghanistan, where it is often called on to curb "unacceptable" attacks driven with almost indiscriminate small arms regardless of the nature of the targets. It is also used to curb escalation of limited conventional weapons clashes, to halt significant fighting in larger conflicts (such as in Syria), and keep former conflicts from breaking out again.

This most widely applied form of deterrence also reflects the rapid growth of collective actor deterrence due to increased efforts to prevent or halt violence and warfare in many places mounted by the UN Security Council, other international organizations, or simply coalitions of willing states. They have never practiced deterrence by punishment, especially randomly, but their efforts tackle some of the most complicated conflicts in the world today. Other factors enhancing attention to deterrence by denial now include:

1) Long-standing concern that deterrence by punishment is, or can readily be, immorally excessive and indiscriminate, thus unacceptable.

2) Facing opponents seemingly impervious to threats (e.g., terrorists, fanatics, a nuclear-armed North Korea) often making threats of punishment and their application underwhelming in their effects.

3) The surge in cyberattacks where opponents are often invisible, undetectable. If they inflict increasing harm and remain very hard to stop states may eventually turn to heavy, even indiscriminate retaliation to get more relief.

4) "Unconventional" warfare in various places, typically as attacks in one state but quite capable of spreading elsewhere.

5) Attackers that will even settle for attacks that just occur without serious harm and destruction.

All this pertains to attacks often not readily prevented and rather costly.

An obvious question: if deterrence by denial is so important, why its slow reemergence lately? One reason is that it normally involves attacks and responses that are familiar and mundane, far from catastrophic, so denial is more widely used. During the Cold War "deterrence" became attached, particularly by armed forces, primarily to threats of nuclear retaliation or major conventional war, even though denial was always widespread. With nuclear deterrence much more recessed now than then, denial stands out more.

Deterrence basically remains the use of threats of harm to prevent harm to you or someone you do not want harmed. Whereas the harm involved was once mostly military in nature, today it takes many other forms too including even nonviolence threats (i.e., the opponent seeks to obtain nuclear weapons or other dangerous ones, or sell such weapons to a hostile third party, or undermine a government one favors, etc.) Deterrence efforts now include sanctions, ceasing arms or aid transfers, harmful financial and trade restrictions—harmful without being violent. The consistent element is feeling one's national security is at stake, perhaps even from cyberattacks of some sort.

As Wilner and Wenger point out in their introduction, deterrence by denial is hardly new but now needs a more sophisticated analysis. For instance, a denial threat normally conveys that it will be too costly or difficult for an opponent to attack successfully, unlike threats of *punishment via retaliation* regardless of how successful the attack is. Denial thus covers: Preparations meant to inhibit future attacks, military or otherwise; Threats to resist harshly—one form of punishment—any attack; Implementation of such threats when necessary. However this makes denial look static, reactive, when it now often involves offensive steps, such as the NATO forces' surge into Bosnia to implement prior threats to intervene in the fighting there to make that fighting look certain to fail. Often such outbreaks may provoke intervention because it is considered by others as disregarding a social contract, violating a set of international norms, or disrupting a system status quo, and the intervention is thus less with punishment in mind than just counteracting the attacker's objectives.

EXAMINING DENIAL IN GREATER DETAIL

Several contributors of this volume highlight Glenn Snyder's 1960 definition of deterrence by denial as the ability to "deny the other party any gains from the move which is to be deterred." A punishment threat clearly aims at an opponent psychologically, steering his decision-making in a compelling way. A denial threat also aims to shape the opponent's decisions, not just psychologically but, if necessary, by physically limiting what the attack would accomplish. The two types of threats are distinct but overlap. Their objective is the same, as is how the deterrence involved should work—frustrating the opponent's plans by threatening unacceptable harm. And as James Wirtz, as well as Janice Gross Stein and Ron Levi point out,[2] both may well provoke a frustrated opponent into seeking ways to design around them. (Designing around denial in dealing with cyberattacks may call for immense effort!) Thus, reviving denial need not fundamentally alter deterrence in theory or practice, conceptually or analytically.

Initially, distinguishing punishment from denial was due mainly to the superpowers. As of the early 1950s, having contemplated catastrophic destruction as their ultimate deterrence threats (as did other early nuclear powers), nuclear powers confronted forces against which there were no adequate defenses. A deterrence failure was potentially nation-killing. Denial was initially seen as a recourse in conventional warfare and thus potentially more tolerable. Warding off attacks by crippling their effectiveness was more tolerable and may have been seen as embodied eventually in missile defense, that was just emerging in primitive versions.

Today, major states' nuclear weapons are more recessed, their relations with each other relatively more relaxed and with the norms of conflict and warfare more limiting in nature, making the differences between denial and punishment less fundamental. This makes distinguishing the two like identifying a difference without any great distinction. Denial simply looks more relevant for our time. However, as noted later, the post–Cold War period has seen the gradual return of the great powers conflicts and the upgrading of their nuclear weapons systems as well as the emergence of increased stocks of the non-great power nuclear weapons states

One adjustment needed as a result now is in detecting potential attacks and attackers hopefully at least in time to prevent an attack, nuclear or conventional, because it can now be harder to both identify and track them. The only comparable problem earlier was that with nuclear proliferation a nuclear war among great powers might someday be instigated by a third party's anonymous attack. It is often harder to both identify and then track potential attackers; it is usually less difficult to identify them after they strike. The problem disappeared for great powers by the missile age, but concern still lingers that, for example, a nuclear power or terrorist might someday succeed in planting a nuclear weapon in a city for later use.

Wilner's chapter emphasizes that earlier in the Cold war there was a stronger tendency to see deterrence as *requiring* threats of punishment.[3] Helping drive this was the expectation that another major conventional war would be devastating and probably escalate to the nuclear level— making denial potentially too costly, especially when Communist bloc leaders were assumed to be indifferent to casualties as long as victory was achieved. The main Western allies thus turned to deterrence by punishment to deter a Soviet nuclear attack and to fear of Western escalation to the nuclear level to deter lesser attacks. Nuclear deterrence as punishment looked attractive for preventing both kinds of war.

As the dominant Western actor, the United States took the lead in pursuing containment in Europe and elsewhere. Western containment was fundamentally deterrence by denial: keeping the USSR and associates from seizing or otherwise adding additional resources to their realm. This had an underlying and ultimately offensive objective: causing the Soviet bloc to decay, fall prey to internal conflicts, lose what revolutionary vigor it had left, and thus break down. Denial was implemented by combined military, economic, psychological, and political means. The military component was labeled "deterrence," but the entire strategy was deterrence, particularly by denial. Critics labeled it too static or inertial, but it was intended to seriously alter the international system. Meanwhile, the Soviet Union's expectation also was that the West would ultimately collapse; capitalism and the "imperialists," held in check enough, would break down militarily and in other ways, altering the international system in its desired fashion.

How did these deterrence postures come to seem dominated by punishment? Starting with the Korean War, the United States and its allies kept large conventional forces close to the communist bloc, to show determination to fight its military expansion, with tactical nuclear weapons stationed nearby. (Soviet forces were similarly arrayed opposite Western forces in Europe.) Even the nuclear forces component had a denial function. Deterrence by punishment with these nuclear forces somewhat in reserve was to be used hopefully on a tactical level only when denial failed, if then.

As denial began to seem too costly in the 1950s in conventional forces and if a conventional war broke out, deterrence by retaliation became a temporary public fixation via the "massive retaliation" strategy. However, this was soon accused of virtually inviting each side to turn any conventional conflict into a race to nuclear catastrophic death and destruction. The eventual US recourse was "flexible response;" use deterrence by denial in a limited war so effectively that escalation beyond that would be unnecessary. But the European allies, nuclear armed or not, objected that even another major conventional war on the continent would be intolerable, envisioning Americans and Russians fighting to the last European. They wanted a short period of conventional war fighting to lead to nuclear attacks on the Soviet bloc—threatening massive punishment to be the true deterrent so no war would result. This resulting two-pronged deterrence lasted out the Cold War: be ready to fight hard enough to show that if necessary, they would resort to vast punishment. The Soviet response combined a huge conventional forces posture up close to the West with threatening that any significant fighting would almost certainly trigger Soviet nuclear attacks to the strategic level—another blend of denial and punishment but with fewer rungs on the ladder.

The result for decades was a string of serious confrontations, limited wars, and dangerous crises (e.g., Korea, Taiwan straits, Vietnam, Berlin crises, Cuba, Afghanistan) involving not only major participants but the use of proxies for both Cold War and other purposes. For example, the United States sought to deter Saddam Hussein (a Soviet ally) from expansionist efforts in the Middle East by asking the Shah of Iran (an ally) to stir up the Kurds in northern Iraq.

As a result, true examples over the years of deterrence strictly or mainly via punishment, even at the conventional level or below and by major powers, were somewhat limited. Some possible examples:

- US Bombing al Qaeda units and facilities in Afghanistan and Sudan for attacking US embassies in East Africa.
- US bombing in North Vietnam retaliating for Viet Cong attacks in South Vietnam early in that war.
- US efforts to disrupt and disperse the Taliban since 9/11.
- US sanctions for years against Iran and Cuba.
- China's invasion of Vietnam.
- Israeli raids into the Gaza Strip.

Other aspects of deterrence by denial also deserve mention. Denial normally involves highly visible preparations because they convey the threat more effectively and may enhance its credibility. Stein and Levi emphasize that denial needs sending clear, crude, and costly signals.[4] Thus a deterrer's capabilities for inflicting denial need displaying to a degree that a capacity for punishment may not. Or from another angle, the capacity for denial is hard to hide, whereas the specific capacity for punishment is often hidden. An example was China's intervention in the Korean War. Beijing threatened denial publicly, as well as privately, and even militarily via some maneuvers in and out of North Korea's border in hopes of achieving denial and showing it had forces ready to use. But it prepared its critical intervention forces and intentions in secret to ultimately launch a huge surprise attack. Failing to deter the United States and South Koreans turned out to be the result, unfortunate because the ultimate price was enormous—so much so that to this day the actual Chinese casualty figures remain secret. The ultimate concern in deterrence by denial is that it may fail in just this fashion, and in the nuclear age this has come close on several occasions.

An obvious question here is: in deterrence by denial, what is most likely to deter? Disclosure of daunting military preparations? (The risk being that an overconfident opponent will attack anyway?) Making it plain the opponent can win only at serious cost—a pyrrhic victory? One might refer to this as raising the price of attacking and dramatizing the potential of a worst-case outcome for the opponent—the risk being that the opponent grasps the scale of the cost only in retrospect. Trying to show the eventual outcome will be bad? Hanoi achieved this in the Vietnam War and won, the United States tried to do the same and lost; the risk was that as sunk costs accumulated the opponent's determination to win rose more and longer.

Thus, a parallel question is how best to sustain deterrence by denial once it is being challenged. Show military superiority at once? But this might help the opponent improve their attack. Display great tolerance for punishment? Maybe, but the USSR did this without it deterring more German attacks or eroding German fighting after the tide was reversed in World War II. Display extreme, persistent intransigence well before and after any attack? That has been the North Korean strategy for decades, helping keep the regime in power but at enormous cost, first during the Korean War and then since.

Remember that deterrence by denial is often conducted after the deterrer is already being attacked. Several chapters—Wilner, for example, like early deterrence analysts did in the Cold War era—emphasizes that deterrence may be very important *after* a war breaks out.[5] It might prevent the attacker from escalating his military effort, discourage a victorious attacker from attacking again, or discourage other possible attackers. As Schelling has suggested, it can also lead to "understandings" with an opponent that contain a war (i.e., putting some types of targets off limits, barring some kinds of weapons, etc.). Dima Adamsky's chapter to this volume emphasizes this is actually fairly common, especially in conducting deterrence over a lengthy time period, like being used to discourage more attacks or punish the opponent for past ones. He explains how Israel has operated this way in recent years, hoping it will over time convince opponents to abandon their attacks via a "cumulative deterrence" effect.[6]

In similar extended conflict situations—like insurrections—repeated attacks by the parties may evolve into a form of communication, expressing determination to not quit or swearing retribution for the opponent's prior behavior, etc. The parties practice deterrence less by attack than by denial via intermittent clashes to show they remain armed and dangerous, will not quit, must be respected, and taken seriously. A common parallel is gang warfare in parts of cities around the world.[7]

On a different tack, missing in most discussions of denial is how it is sought via preemptive attacks to destroy or weaken the opponent's attack capabilities, something Wirtz points out in his contribution.[8] The omission of this is surprising in view of US and Israeli interest for some time in possibly attacking Iran's nuclear weapons facilities, and Israel's frequent raids over the years to prune its other opponents' offensive military capabilities. If the aim is clearly and solely to avoid a future attack by the target actor this is roughly the same as denial by defensively defeating or degrading his forces—denial by degrading in advance. This widens the concept of deterrence somewhat, complicating some conflict situations. Thus, during the Missile Crisis the United States threatened, and began planning for, mounting a preemptive attack to prevent any missile attack from Cuba, to also punish Moscow for sending its missiles there, and to deter it from sending any more. This looks mainly like denial, but it was also partly a punishment threat as well—Kennedy threatening that any use of the missiles in the hemisphere would justify "a full retaliatory response" on the USSR. The United States has more recently practiced denial, at least in part, of future terrorist attacks by drone assassinations of terrorist leaders—retaliatory deterrence at once via punishing attackers, doing harm to some noncombatants, plus denial by disrupting plans for future attacks.

DETERRENCE BY DENIAL VIS-À-VIS A PANOPLY OF NEW THREATS

Can analysis of deterrence by denial be appropriately employed along roughly the same lines in dealing with very different contemporary problems: terrorism, international threats to peace and security, and cyberattacks? Or do such threats differ enough to require different treatment? Are these threats unique? For instance, modern deterrence, in theory and practice, has mainly envisioned highly organized actors, particularly states, as the parties to be deterred, contained, or defeated. But it is hard to readily link cyberattacks to specific states consistently— some may leave a return address, but most do not. (It is unclear how many reliable countermeasures are unused, leaving the relevant full deterrence measures unknown or ambiguous to attackers.) Terrorism has some links to states, but it typically comes from semi-autonomous or private sources, weak states, or transnational entities. And internal threats to broader peace and security—insurgencies, genocides, outright civil warfare—typically arise in states with marginal cohesion and rule.

Terrorism threats have evoked extensive responses heavily oriented toward denial, including preemption. They remain relatively uncommon though hardly rare. But cyberattacks are now ubiquitous. Many states and societies are barraged daily with denial efforts; the main, often the only, response, and the problem is steadily growing.[9] This kind of harm is widespread, of uneven impact, quite disturbing, and widely practiced and is often very difficult to effectively counterattack, especially with highly effective efficiency, and thus is readily handled by denial practices when possible. In fact, denial is often the only response. Some of the most serious cyberattacks thus far have been from states (such as Russia) or well-organized groups. Cyberattacks should potentially be capable eventually of posing even serious military threats, and already do serious harm in intelligence or inside existing cyber systems.

The threats vary significantly, as do the attackers. Terrorism is often practiced by people with very unusual objectives, often with rather primitive means. Cyberattacks come via advanced technical capabilities employed by very unusual people that are certainly not primitive. Terrorism is only marginally linked to the advanced systems, state sponsorships, and state support often underlying cyberattacks. We treat terrorist attacks as dire and the perpetrators as minimally deterrable; we use denial because they are often impervious to harm or death threats and unmoved by fear of failure. Denial is employed mainly for preventing them from being in position to attack rather than convincing them not to bother—deterrence by threat of interdiction that verges on not being deterrence at all.

Terrorism and cyberattacks seem worlds apart from weak states with ferociously harmful regimes or highly radical opponents. Modern deterrence emerged to assist states in fending off attacks by each other, but today it is often used even to ward off unacceptable behavior, even if it is not particularly aimed at us. That behavior might, for instance, open doors for thefts of nuclear weapons or provide conditions that facilitate setting up bases for terrorists' operations, while not threatening other actors, at least not immediately. The threats are not normal attacks but rather violations of human rights, massive refugee movements, ubiquitous corruption, brutal repression, and the like. Deterring these "attacks" departs starkly from deterrence as traditionally conceived and practiced. The perpetrators do not attack so much as just disgust us into trying to do something about them.

Comparing these kinds of threats with others on how they are handled starts by noting that the necessary cooperation varies a good deal. It has been fairly, even very, intense toward terrorists, particularly on sharing information on threats, developing detailed tracking and screening systems, and exchanging advice. However, threats that violate many international rules draw rather uneven cooperation for suppressing them, up to and including clashes in the UN Security Council. On cyberattacks, seeking private aid is more feasible than with the other threats, increasing chances of success but only partially. Governments do not readily share all they know, and private sector victims are often reluctant to do so either.

Thus, lumping these threats together intellectually and for trying to defeat them, aiming at careful analysis of the nature and role of the deterrence involved, invites skepticism about its utility in the end. Deterrence of terrorists seems to have evolved into a relatively cooperative and sometimes integrated effort, particularly for nabbing perpetrators and disrupting attacks in advance. However, cyberattacks involve mainly seizing valuable intelligence and related information and there is little or no international law barring that. And while major states are often the most targeted, they are also in many cases major perpetrators. Deterring gross violations requires threats with real teeth but often the necessary cooperation cannot be arranged politically, either internationally or at home. And the threats and related forces mounted are often too limited or restricted in use, too poorly trained and equipped, or lacking in credibility to handle the situation.

SOME ADVANTAGES OF DENIAL

Deterrence by denial has its advantages and disadvantages that require further analysis. A nice introduction is James Wirtz's assertion that denial often allows the practitioner to control its success.[10] A standard concern is that this deterrence requires acquiescence from the opponent to succeed—the opponent controls that. That remains true when the deterrer is threatening to deny the opponent a successful attack to prevent it from occurring (like the North Korean nuclear weapons program and its acquired nuclear attack capability that resulted despite US and ROK efforts to prevent it). Once an attack is launched, success in denial is then normally in the deterrer's hands by defeating the opponent to complete deterrence success.

One concern bearing on this is the credibility problem, particularly in practicing extended deterrence. To dictate the outcome and thus deter requires the necessary credibility. Without that what can happen is that the opponent underestimates the deterrer's strength; or doesn't believe the deter will use it; or believes the opponent will tire faster in a fight and it can therefore outlast the deterrer; or wants to risk defeat to make a point anyway; make a recurring effort to wear the deterrer down; or call attention to his cause for seeking bigger and better support next time. This is often not an attractive outcome for the deterrer. There is also the problem of carrying a victory too far, too viciously, or too harmfully, resulting in the frustrated opponent eventually turning to do almost anything that will be harmful, turning a seeming victory to a lengthy violent resistance, all too common in the Middle East and Southeast Asia. Thus, figuring out how to detect and deal effectively with such situations would be wise for effectively enhancing the appeal of many deterrence-by-denial efforts.

The next advantage of denial deterrence is highly appealing because it can perhaps offer greater situational clarity. Several chapters in this book emphasize the need in deterrence by denial for offering clear and credible threats, readily perceived and understood as defensive in nature.

But deterrence by punishment may well require that the deterrer hide some of his specific plans and capabilities lest they incite a preemptive attack (a maneuver states such as China and Pakistan have employed) or to avoid a security dilemma which provokes an attack deterrence is supposed to avoid. In the Cold War era, for example, trying to construct even a limited ballistic missile defense (BMD) was readily interpreted by the opposition as making preparations to eventually attack, and it is still regarded this way in Moscow and Beijing. And annual US–ROK military exercises have regularly driven Pyongyang into both a frenzy of expensive warlike preparations for fear they are cover for a surprise attack, and thus an explosion of objections.

Another advantage of deterrence by denial, when carefully utilized, is how the defender can extract visible information on how his deterrence is doing from the opponent's reactions. Major improvements in a defense posture can provide valuable information when the reaction is a parallel defensive or offensive buildup, or turning toward a new military posture, or making no response at all. Jonathan Trexel notes, for example, that BMD might dissuade opponents from seeking to obtain or develop missiles, or BMD might enhance a potential attacker's uncertainties about whether an attack would be a good idea. He adds that BMD can be reassuring about otherwise fearing the opponent could be becoming irrational or planning to attack, because it at least offers promise of successfully coping with his attack.[11] (Of course, this can readily undermine a mutual deterrence relationship if it is not carefully prepared.)

Next, denial is inherently defensively oriented, with capabilities meant to frustrate and defeat only attacking military forces and capabilities, such as at an emerging point of potential attack by another party, or by only ever aiming at opposing forces that arrive in one's neighborhood. Thus, a denial threat can be more normatively acceptable under such circumstances, something useful politically with and for soliciting support from others. To a third party being defensive in orientation is normally much more likely to look acceptable than looking attack oriented.

However, exceptions can seriously complicate matters. For instance, when denial frustrates attacks aimed at regaining territory the defender earlier seized, observers may side with the attacker. This has been true for years in the Israeli-Palestinian conflict and conceivably may be true soon regarding islands in the East and South China Seas. Denial may also be normatively unacceptable in interstate conflicts when ethnic or religious groups want to control or defend or expand their territory, particularly if the state disintegrates, something that can readily emerge, and which sometimes leads to genocidal results.

It is also important to note that, in line with prospect theory, a denial posture may well convey more emotional and political commitment or staying power to fight because a loss or set of losses (and not some seizure of gains) is at stake for the deterrer. Therefore, it may also simultaneously encourage the opponent to see withdrawing from his planned attack as less risky and worth considering. A denial posture is often condemned as a fraud in such intense conflict situations, so much so that external forces must be inserted, as by a collective actor, to resolve that credibility problem.

However, denial can therefore facilitate discarding punishment deterrence postures and capabilities. This would be pertinent now if any discussions were to emerge about eliminating nuclear weapons, because enhanced denial capabilities would probably be necessary if a number of states were to accept and implement the necessary steps. For instance, having reduced their ability to defend themselves since the Cold War, Russia, Britain, and France are relying more heavily on deterrence by punishment as their ultimate recourse for protection, at least for the time being. India and Pakistan are in the same situation. For a number of states, nuclear weapons abolition may become acceptable only if they turn to reliance on upgraded-denial nonnuclear postures again.

That deterrence by denial can facilitate management of an important sector of international politics also has much larger implications. Deterrence, particularly denial in the contemporary international system, is crucial to the maintenance and conduct of overall system management. This is the case when it comes to suppressing WMD proliferation, the spread of violent unrest in a regional system, or the overall maintenance of order. This is where a new and penetrating analysis of deterrence is now needed the most.

A good example is how the major deterrence target today is internal fighting, to a greater extent in fact than interstate conflict, particularly as it poses major ongoing or potential problems for particular states and societies. Such situations, when rather serious, should normally be handled with very limited recourse to punishment per se, particularly when overseen by or mounted under the auspices of international organizations. IOs are normally in a poor position to effectively impose deterrence by denial on their own (a good example has been in the ongoing war in Syria). They have to mobilize an international collective to mount and enforce threats, and are often unable to put forces in place and take other denial steps well in advance—members seldom fully agree on doing this (also frequently displayed in Syria). As a result, denial requires using at least some initial offensive steps to squelch the fighting in hopes of then keeping it from returning. Stability may eventually require even crippling one or more parties. This is a particularly difficult sort of denial, as evidenced in the situation in Afghanistan for almost two decades.

DISADVANTAGES OF DETERRENCE BY DENIAL

In deterrence there never seem to be advantages unaccompanied by disadvantages. One example is that, as suggested earlier, deterrence by denial can be quite complicated, particularly because of sharp and surprising developments in both international and domestic affairs, particularly conflicts. For instance, for using punishment in deterrence there is no inherent reason it fit the nature of the attack—just that it be able to punish or threaten to punish, in ways and to the extent necessary, to sufficiently hurt the opponent. Denial must threaten to turn the opponent's attack itself into a failure, and then, if the attack comes, to carry out that threat. This therefore includes accurately ascertaining or estimating both the nature and scale of the attack, and especially the attacker's motivation. Mistakes about this naturally occur in many cases, before or during the attack. Overestimating the threat and preparing accordingly can help a good deal in offsetting and coping with the mistakes but only with a larger expense as well, and over time the deterrence effort may erode as a result of having to sustain the burdens involved.

Next, being obvious about plans to cope with an attack may unfortunately provide key information an opponent can utilize to facilitate his attack, designing around the opposing deterrence posture with more knowledge and confidence. Unfortunately, though the best defense may require keeping some secrets, a deterrence posture requires being advertised and perceived in a suitably threatening and credible way. Bluffing can work but if detected not only might it incite an attack but give the deterrer an unhealthy reputation in the future that damages its credibility. On the other had being more open about one's capabilities can make it harder to bluff successfully when they are indeed inadequate. All this can make denial expensive to prepare, sustain, and keep safe from probes.

Another factor is that it may be harder to bluff in a deterrence-by-denial effort because denial postures can often be less dangerously probed for vulnerabilities. Several authors in this book (Stein, Levi, Wirtz) point out this can incite an opponent's efforts to design around them. That can mean having to prepare to defend against a wide array of contingencies in multiple ways, adding significant expense especially in maintaining backup capabilities in case things go awry. This was one reason that early in the Cold War the United States and the West decided it would probably be too expensive to do denial only. The Soviet Union eventually poured vast resources into this sort of deterrence posture, and by Mikhail Gorbachev's time this had become ruinous indeed. Since the Cold War, the United States has been repeatedly disappointed by allied unwillingness to sustain military capabilities suitable for forceful interventions or threats of them on behalf of sustaining international peace and security. Of course, the unwillingness is easy enough to explain.

A final element in conducting deterrence is paying close attention to the strength of the opponent's motivation, which is always crucial. Generally, the greater that motivation the greater the need for constantly upgrading one's denial capabilities, the longer the threat is likely to persist, and the more varied it may be as the opponent tries various options. The worst threat situation is probably when the opponent's motivation is strong and the costs of attacking are fairly, even tolerably, low—such as in the opponent undertaking cyberattacks. A good example of all this is the numerous times Israel has been attacked on a limited scale over many years and will almost certainly continue being attacked in the future.

Alongside this, in even more closely examining deterrence by denial, it seems that attention to the perspective of the deterrer, who might well become an attacker if suitably provoked, is also required. Often sustaining deterrence between states and responding to grievously unacceptable behavior within a particular state (or states) requires not being deterred from mounting attacks to prevent or halt the harm being done. In effect, this is shifting the tables—the deterrer becomes the attacker. Concern about this has been a long standing component of deterrence, for example the way the United States and others have feared nuclear proliferation—worrying that if a state like North Korea or Iran, or possibly al Qaeda or ISIS gets a nuclear weapon then it would potentially become impervious to normal complaints or threats. Under these circumstances deterrence by denial can be intensely valuable for forestalling an opponent's breakthrough to a new attack/deterrence threat.

CONTEMPORARY DEVELOPMENTS AFFECTING DETERRENCE BY DENIAL

Western deterrence threats and use of force to repair, sustain, and even improve national and international security have been facing very serious difficulties for some time that have been slowly but steadily expanding. An important initial factor has been that many Western governments have had little domestic support available for such burdens—"why bear the costs and casualties?" is the reaction. In the liberal international world this has been further reinforced by the emergence of huge uneasiness about, and often serious opposition to, immigration from third world nations and even nearby Western nations, adding additional costs and burdens. The initial drive behind all this was the disappearance of the Cold War and not long after that the Great Recession—those developments made it no longer seem necessary and useful to uphold the elaborate support that had been sustaining the dominance of the Western world for decades.

The resulting decline of liberal internationalism and the collective support that had long sustained it has been the most important factor. It is much more difficult now to pull the West together for a serious use of force, or to mount serious threats of force, in many parts of the world. A major additional factor has been a result of the failure of the West to absorb Eastern Europe and Russia into the West after the collapse of the Cold War and the Soviet Union. The recession was, of course, very hard on nearly everyone but much more significant in international security matters has been the failure of Eastern and Western states in the newly emerging liberal international world to successfully merge smoothly, which led to the return, over the next nearly three decades, of autocratic rule again in the East, especially in Russia. This has meant the return of large Russian forces being poised along its borders and aimed largely at the West, a sharp breakdown in relations between Russia and the West, and the movement toward closer connections with Russia by Poland, Hungary, and Romania, among others. In Europe much of the remaining effort to sustain security has been a return to costly deterrence postures but with little enthusiasm. An American-Russian nuclear weapons competition has returned with the resulting abandonment of certain aspects of prior nuclear arms agreements. The period when wars and near wars, interstate and intrastate, were noticeably declining has given way to a period in which international security management is slipping away, turning the East-West relationship into a semi Cold War. Added to this has been a strong surge by China to greatly strengthen its place in the international system via demanding that a good deal of the areas around it are supposed to be China's or at least something like China vassals. Included in this is a strong Chinese nuclear weapons program imitating many American weapons, accompanied by a parallel surge in Chinese conventional forces and considerable theft of Western advanced Western technical knowledge in general—in short, a further addition to the semi-cold war.

However, this deepening of the need for deterrence once again is taking place amid sharp alterations in the international security situation of a different sort. It is the result of an enormous explosion in technology which will almost inevitably alter the world and international affairs via the resulting changes. It is becoming much more difficult for countries to protect themselves from all of this, including its anticipated impact on national security via major alternations in military affairs. As a result, it has already started making it more difficult for countries to protect themselves from ever more sophisticated weapons that are emerging— for example, much more accurate missiles and destructive cyberweapons guided by outer-space satellites. Consequently, chances of exposure to severe disruption or destruction, even without nuclear weapons attacks, are steadily rising, particularly from advanced technology developments.

Deterrence is therefore steadily becoming more difficult to count on for 1) successfully fending off attacks, 2) successfully inflicting attacks, and 3) successfully ever doing away with nuclear weapons, other severe weapons of destruction, and other disturbing threats. Deterrence itself may well slip away as the ultimate recourse for sustaining international security in the foreseeable future because artificial intelligence (AI) has begun to strip away places and ways to hide. It may also occur because cyberattack capabilities are emerging that will be able to disrupt defenses quite readily, even the key centers for the conduct of national military operations. Adding to all this is the rapidly growing expansion of AI, including the huge increase in monitoring almost every portion of the earth from the myriads of satellites and other measuring and detecting systems multiplying in outer space which is already taking place; in short, a growing capacity to track almost everything.

These and related developments will readily give the major powers considerably greater deterrence capabilities against most states, retaining for them the utility of deterrence by denial in much of international affairs. However, it will confront those major powers with having potentially much less reliable deterrence capability among themselves, less reliable in being consistently subject to a possible major attack on very short notice in their relations among each other.

Here are some of its major components at work now. Artificial intelligence is exploding in many fields, especially military and security affairs where vulnerability from transparency is growing via emerging commercial surveillance satellites, drones, smartphones, computers. Monitoring and tracking military operations, tracking missile firing preparations and releases, military communications, etc. are massively expanding. Cyberattacks, as is well known, are now ubiquitous as are the expanding efforts to detect and dismantle them.[12] China has announced plans to spend billions of dollars on AI research and development over the next two decades.[13] Studies now strongly assert that AI will dramatically alter many aspects of national security affairs—military, economic, and informational—within the next several decades,[14] and they anticipate that this will radically transform these fields. Apparently the already-huge collections of AI data will be readily superseded, and there will be much enhanced forgery in audio and video media operations, plus what will amount to a new industrial revolution as well as a huge impact in nuclear, cyber, aerospace and biotech areas. As a result, there will be a much larger role for cyberattacks in military offense and defense activities as well, such as in disrupting command and control systems, and better targeting of opponents' forces and facilities.

Adding to this list is the sharp rise in additive manufacturing, more commonly referred to as 3-D printing, which is already being widely used in aerospace and other defense industries. It involves using a digital process to generate components one layer after another which then allows the production of complex products with elaborate shapes by machines using automatic processes rather than skilled workers, all at far greater speeds and lower costs. Really an extension of AI, it is a process that can all too readily be copied, and thus also stolen or otherwise secretly transferred to others by hackers. It is also therefore vulnerable to the reverse—a hacker or other process transferring information and instructions that leads to crippling or sabotaging a 3-D process. The obvious targets would be someone else's military weapons and transportation systems, nuclear weapons and their manufacture, the disruption of the equipment and information associated with those activities, and the weakest links in those activities. So readily is the transfer of these practices operable, an advanced version of 3-D printing pertaining to military activities is now referred to as an "internet of nuclear things," layers of technology tied together by digital threads."[15]

All this suggests that the arena for the conduct of deterrence by denial will be subject to considerable alteration, even some upheaval. There will be greatly expanded capabilities, huge new resources, far less secrecy, and perhaps many more vulnerabilities, presumably leading to significant alterations in how to mount and conduct various versions of deterrence. There is already a considerable surge in efforts to lead, match, or catch up to these developments across the Western world, Russia and China, South Korea and Japan, India, and numerous other states. Many of the new resources are already being stolen

As a result, one must anticipate a very different deterrence environment, one that is very dense and complex because states will be so much more intersected, more interpenetrated, and potentially very much more dangerous.

CONCLUSION

Deterrence by denial has deep roots and a long history. But clearly much more can be done in shaping its design and conduct to make it more effective, and things will have to be done to adjust to the upcoming changes in threatened and violent international conflicts. This book offers a variety of perspectives, predictions, and solutions that can help in further refining deterrence by denial in theory and practice. The study of deterrence by denial will continue driving us to bring out our best to cope with some of the worst in others. It will remain a vital recourse for the better management of peace and security, nationally and globally—the spine of an effective international security management system. Doing justice to those who initially led the way in designing this endeavor, in what seems so long ago and far away, requires no less in the way we mount our efforts now to refine the theory and analysis of it in the years ahead.

Notes

1. See Introduction to this volume.
2. See chapters 5 and 3 in this volume.
3. See chapter 2.
4. See chapter 3.
5. See chapter 2.
6. See chapter 7.
7. This point is further developed in Lieberman, *Reconceptualizing Deterrence.*
8. See chapter 5.
9. On the cyberattacks problem see, for example, Buchanan, *The Cybersecurity Dilemma*; Libicki, *Cyberspace in Peace and War*; and Hennessey, "Deterring Cyberattacks," 39–46.
10. See chapter 5 in this volume.
11. See chapter 6 in this volume.
12. Larkin, "The Age of Transparency," 136–146
13. Metz, "China's Blitz to Dominate A.I."
14. Allen and Chan, "Artificial Intelligence and National Security."
15. Hoffman and Volpe, "Internet of Nuclear Things," 102–113.

Chapter 2

Dawn of a New Deterrence

Intra-conflict, Cumulative, and Communicative Denial

Alex Wilner

Deterrence skepticism runs deep: The fall of the Berlin Wall, and later, the felling of the Twin Towers in New York City, forced some scholars and decision-makers to question its continued utility. To some, deterrence theory's core concepts were deemed obsolete and tired, altogether unable to address emerging contemporary security challenges.[1] Other guiding principles, like preemption, were championed in its stead. But deterrence theory is not easily superseded or altogether jettisoned. Instead, its tenets can be re-assessed, in some cases broadened, and expanded to reflect global developments and evolving insecurities. If Cold War deterrence theories were a reflection of the particularities of the Cold War era, then contemporary deterrence theories should seek to take into account the shifting and evolving security landscape in which the theory evolves. And as Malcolm Chalmers further reminds us, deterrence "should be seen … as one of several tools for the prevention [and] … management of war," sitting alongside other relevant concepts, like compellence, influence, coercion, and assurance, as Morgan, Wirtz, Sawyer, and others illustrate in this volume.[2] The effectiveness of some specific deterrence concepts and strategies (like mutual assured destruction) may ebb and flow, but the paradigm itself, and deterrence theory's core objective of managing conflict writ large, are not so easily discarded.

In some respects, the international security parameters within which deterrence theory first emerged have given way to something new, more fluid, and complex. It may be cliché to argue that the world keeps on changing, but it also happens to be true. Deterrence theory thus must adapt and respond in kind. Guided by Cold War parameters, traditional deterrence theory largely focused on state and government actors, interstate conflict and war, and high-octane threats of punishment, like nuclear escalation and retaliation. Today's parameters, conversely, provide a wider array of concerns and challenges; whereas states are still in the picture, a variety of violent substate and non-state challengers have crowded the frame. Conflict within states (civil war, insurgency, rebellion) and between and among disparate actors (terrorism, cyberwar, piracy) have gained in prevalence vis-à-vis conflict between states. And whereas nuclear proliferation and the threat of nuclear warfare still loom large—especially among great nations and their rising competitors—conventional conflict demands ever more attention. Under these conditions, repeated and ongoing serial engagements and iterated confrontations between states and non-state adversaries (including militants) are possible and perhaps even likely. The conflict with the Islamic State, al Qaeda, and their supporters, for instance, has been dubbed the Long War, a conflict with no clear or certain end. Rather than assume that deterrence theory is ill-equipped to deal with the threat of continual, low-intensity conflict, one should explore how these campaigns of violence might be met with campaigns of deterrence.

Importantly, as Cold War conditions have receded, deterrence by denial has been provided with new ground to explore. Deterrence during the nuclear standoff largely took place at the strategic level, in which strategic stability between the United States and the USSR, Morgan suggests in his contribution, required that both sides relinquish a larger role for denial.[3] Strategic stability stemmed from purposeful vulnerability; mutual and assured pain of punishment. But now, and under conditions of conventional conflict more broadly, deterrence is increasingly about practicing denial. Non-state, asymmetric, and cyber-based adversaries cannot easily encourage or expect strategic level stability with their much larger and more powerful state adversaries. Nor will their deterrent relationship every truly entail mutual vulnerability. For example, awful as terrorism is, al Qaeda and IS are a nuisance, not an existential threat to the United States. Herein, defenders are free and more capable to use the logic of denial in deterring violence. In serial engagements, as Dima Adamsky and Jonathan Trexel highlight in their respective case studies of Israel and Japan, and as Martin Libicki suggests with regards to cyberspace, practicing denial may make a lot of sense.[4]

This chapter updates contemporary deterrence by denial by first reaching back into traditional deterrence theory to reconceptualize certain coercive concepts, and then by reinserting these concepts into our existing thinking of contemporary deterrence. Three concepts are explored in particular: *intra-conflict* deterrence by denial; *cumulative* denial; and *communicative* denial. Descriptive scenarios and contemporary vignettes are used to help illustrate each. By broadening our understanding of denial, this chapter suggests ways in which deterrence might be appropriately applied to contend with contemporary security issues.

INTRA-CONFLICT DENIAL

Deterrence does not end once a conflict begins. True, deterrence is primarily about avoiding war and persuading adversaries from initiating attacks. But deterrence theory does not disappear once a crisis tumbles into open conflict. Rather it skips into a secondary realm: deterring, compelling, and/or manipulating behavior within war. The logic of intra-war (or, in this case, intra-conflict) deterrence emphasizes that adversaries can use and communicate coercive threats while concurrently carrying out military operations against one another.[5] It is deterrence during, rather than before, a conflict. Its premise is based, at least partially, on the idea that deterrence (and coercion more broadly) is practiced as a campaign, one that stretches out over time. Deterrence evolves in response to shifts in the security landscape. Accordingly, it begins in peacetime, where the threat of military engagement is deterred, but continues throughout subsequent phases of crisis and conflict. "A campaign approach to deterrence … is necessary," write Kevin Chilton and Greg Weaver, because as a conflict unfolds "the content and character of a foreign leadership's decision calculus can change significantly." What might have deterred an opponent at one point during a period of tranquility, or crisis, or conflict, may not hold sway at another point, as circumstances, goals, and capabilities change.[6]

Intra-conflict deterrence, then, involves influencing the way a war is conducted; its goal is to shape adversaries' behavior by placing limits on the scope, nature, and/or ferocity of a conflict. Other concepts, like *narrow* deterrence, are closely related. Lawrence Freedman explains that "narrow deterrence involves deterring a particular type of military operation within a war" (as opposed to *broad* deterrence which deters all war, more generally).[7] One way to relate these concepts is to consider intra-war deterrence the macro-concept (i.e., deterrence within war, writ large) that subsumes narrow deterrence within it (i.e., deterrence of specific actions, within a war).

During the Cold War, intra-war deterrence largely addressed the issue of escalation and sought to provide warring parties with opportunities for mutual restraint, curtailing their choice of targets and weapons.[8] For instance, if conventional war between the US and Soviet blocs had broken out in Europe or the Middle East, intra-war deterrence suggested strategies both parties could use to help ensure the conflict was waged within limits (i.e., did not escalate to nuclear exchange). But if nuclear weapons were indeed used in Continental Europe or the Middle East, intra-war deterrence sought ways to deter the further use of these weapons against US and Soviet homelands. These scenarios are reminiscent of Herman Kahn's distinction between vertical and horizontal escalation. In the former case, a conflict expands upwards, towards more lethal and more brutal forms of violence; in the latter, a conflict spreads outwards, engulfing new territories and belligerents.[9] Intra-war deterrence can involve forestalling either or both eventualities. Of course, the concept is not without its faults or inherent difficulties. Communicating deterrent expectations within a conflict, and with an advisory one is shooting at no less, can be tricky; miscommunication may be at a premium.[10] And if decisively winning a conflict is a strategic goal (or an existential must), an adversary may be hard pressed to accept restraints on its behavior in any intra-war coercive bargain. These and other dilemmas will mark how intra-war deterrence is put into practice and may dictate its success and failure.

A classic historical example of intra-war and narrow deterrence—that has been tested in Syria more recently—is the use of chemical weapons on the battlefield. As Freedman and others illustrate, the development, weaponization, and wide-spread use of chemical weapons in World War I forced combatants to find ways to deter their further use on future battlefields.[11] This intra-war deterrent goal has largely succeeded across time and spectrum; in the century following WWI, only a handful of wars and conflict involved the use of chemical weapons.[12] Even Nazi Germany, which had no qualms using gas to kill millions of Jews and others during the Holocaust, was deterred from using chemical agents against the Allies, even on the eve of its capitulation. More recently, in the lead-up to the 1990–1991 Gulf War, the United States and its allies were especially concerned that Iraq President Saddam Hussein would order the use of chemical weapons against Israel, Saudi Arabia, and the United States, and Coalition troops stationed in the region in response to US efforts to liberate Kuwait from Iraq. Baghdad had previously used gas in its eight-year war with Iran and in attacks against Iraqi citizens of Kurdish descent. How serious was the American concern? During the crises and ensuing conflict, US President George H.W. Bush was accompanied by a military officer who carried a gas mask in case one was needed.[13]

How might the logic of intra-conflict deterrence be intertwined with the logic of deterrence by denial? Herein, threats of failure would be used instead of (or along inside) threats of punishment to delimit an adversaries' behavior and/or to manage the way a conflict is pursued or evolves. At the tactical level, warring parties might be able to communicate an ability to curtail the desired effectiveness of certain weapons systems or deny opponents the opportunity to attack certain targets. At the strategic level, an adversary could be denied the goals it seeks by acting in particular ways. And returning to the notion of a deterrence campaign, denial strategies could evolve over time, responding to the particularities of the crisis or conflict. Importantly, the potency and efficacy of specific intra-war denial mechanisms might be less affected by shifting environmental (i.e., military) conditions. Deterrence by denial may change less dramatically over time and over the development of a crisis or conflict because denial is a result of what an actor has (i.e., defence), whereas punishment is based on what it can promise to do (i.e., retaliate).

As illustration, consider Iraq's non-use of chemical weapons (CW) during the Gulf War.[14] One compelling view is that Hussein was deterred from launching chemical attacks against US forces stationed in Saudi Arabia and against Israel because of credible threats the United States (and Israel) had communicated prior and during the military engagement. President Bush, for instance, on January 5, 1991, addressed a letter to President Hussein (which was personally delivered by Secretary of State James Baker to Iraq's Foreign Minister, Tariq Azizi) stating, bluntly: "...the United States will not tolerate the use of chemical or biological weapons... The American people would demand the strongest possible response. You and your country will pay a terrible price if you order unconscionable acts of this sort. I write this letter not to threaten, but to inform."[15] Secretary Baker later restated the warning: "If the conflict involves your use of chemical or biological weapons against our forces, the American people will demand vengeance. We have the means to exact it. ... this is not a threat, it is a promise."[16] US coercive threats continued after Operation Desert Storm began. Weeks into the campaign, Defense Secretary Dick Cheney reiterated: "I assume [President Hussein] knows that if he were to resort to chemical weapons, that would be an escalation to weapons of mass destruction and that the possibility would then exist, certainly with respect to the Israelis ... that they might retaliate with unconventional weapons as well."[17] There was enough ambiguity in these and other threats suggesting that the United States or its allies might resort to using nuclear weapons against Iraq in the event it launched chemical attacks. Such threats may have swayed President Hussein's calculation. Thus, while Iraq did launch nearly 100 Scud missiles against Israel and Saudi Arabia in the hours following the commencement of hostilities on January 17, 1991, not one missile contained any chemical agents.[18]

Missing from this exploration of US intra-war deterrence with Iraq is the effect intra-war denial may have played in shaping President Hussein's decisions to refrain from using chemical weapons. In planning the war, US decision-makers were exceptionally wary of Iraq's chemical and biological capabilities, notwithstanding Hussein's penchant for using WMD in battle, and took appropriate measures to deny Baghdad the ability to use such weapons to its advantage. In preparing for Operation Desert Storm, the United States set up over sixty hospitals in the vicinity, assembled several thousand sickbeds in the war zone, and had at least two hospital ships on hand in the region.[19] US soldiers were also administered pyridostigmine bromide, a pretreatment prophylactic meant to curb the effects of certain chemical nerve agents. These and other preparations were an attempt to help neutralize the effect of Iraq's chemical and biological agents, had President Hussein ordered their use in battle (or presumably, in the case US strikes inadvertently led to their accidental release). Likewise, during the Cold War the United States and its NATO allies had prepared to face and fight the Soviets on a chemical-laced battlefield. Much of that capability—in terms of training, preparation, and protective gear—was in play during the Gulf War.

Thus whereas Iraq may have had some confidence regarding its ability to effectively launch CW offensively (after all, it had extensive practice doing so), it might nonetheless have doubted its ability to actually inflict harm on US soldiers and/or on its ability to deploy an effective fighting force under chemical conditions against better prepared Coalition soldiers. "US forces passive defenses," suggests Barry Schneider, might have "played a major part in Iraq's decision not to use chemical arms, perhaps as great a role as President Bush's implied nuclear threat."[20] American and Coalition forces were prepared to absorb and recuperate from chemical attacks, and to then fight under chemical conditions. This was intra-war denial: denying Iraq the gains it sought with chemical strikes. That preparation was done in the open, reassuring US soldiers (and their families and the American public more broadly) that Iraq's chemical weapons would not greatly alter the course of the conflict and could be neutralized, and communicating a denial capability to Baghdad that illustrated its inability to gain what it sought by using CW in battle. In sum, Iraq might not have been sure that chemical attacks would have achieved their tactical intention—the death of thousands of US troops in Saudi Arabia and Kuwait and a stalled coalition offensive—or strategic purpose—turning Western public appetite against the war and strengthening anti-war attitudes in Washington, DC.[21]

CUMULATIVE DENIAL

A subset of contemporary deterrence scholarship explores whether and how prior deterrence successes and military victories might compound over time, shaping future coercive scenarios. This area of research taps into the broader discussion of reputation in international relations but focuses more on exploring how iterated interactions influence the practice and outcome of coercion.[22] For illustration, in his contribution to this volume, Adamsky argues that in practice Israeli deterrence turns into a series of "forceful acts aimed at educating a challenger about the 'rules of the game,' and to force them ... 'to internalize these lessons' over an extended period of time."[23] Cumulative deterrence suggests that successes can pile up, one atop the other, such that the end result is an eventual and fundamental shift in adversary behavior. Uri Bar-Joseph is usually credited with coining the term. Using Israel as illustration, he explains: "*Cumulative deterrence* is the long-term policy which aims at convincing the Arab side that ending the conflict by destroying the Jewish state is either impossible or involves costs and risks which exceed the expected benefits."[24] The goal is to figuratively bank Israel's prior history of militarily defeating its opponents, and to add Israel's known and suspected conventional and nuclear military superiority to the mix, in order to persuade adversaries that further violence against Israelis is altogether futile. Conceptually it combines elements of punishment ("look how strong we are") with denial ("look how weak you are; you'll never win, so quit trying"). More recently, some scholars, like Doron Almog, Paul Davis, and Shmuel Bar, have applied the concept to deterring terrorism and non-state violent actors more specifically. Thomas Rid has done so as well, though from a criminological perspective.[25]

Cumulative deterrence has been widely, and at times convincingly, criticized. Martha Crenshaw argues that it is unclear "how continuous pressure and use of force can be defined as deterrence." Deterrence by punishment is based on threats, not repeated action. "The defender," she concludes "does not say 'we will keep hitting you over the head with this hammer so you don't even think of attacking us' but rather 'we are not hitting you now but will hit you really hard later if you cross this line'."[26] Janice Gross Stein's reservation is equally damning: the concept "so badly violates the fundamental meaning of deterrence that it loses any analytical value. That kind of strategy is better described as—war."[27] And, more broadly, the outcome cumulative deterrence aspires to achieve— an end to conflict between antagonistic sides—appears similar to general deterrence. In general deterrence adversaries who retain the capability to harm one another do not, and instead use the open knowledge of their capabilities (along with vague threats) to pacify and police their relationship. Some conceptual overlap may be inevitable and is certainly acceptable, but in this case cumulative deterrence looks a little bit like a different shade of black.

And yet, in relation to counterterrorism in particular, the concept of cumulative deterrence by punishment appears to hold some sway. Deterring terrorism uses flexible goalposts. Cold War notions of deterrence success—which were largely absolute in nature (either you deterred an opponent or you did not)—and deterrence failure—which were tallied once threats of force crossed over into the use of force—are less evident. Instead, deterring terrorism often includes partial success, at the fringes of adversarial behavior.[28] John Sawyer explores this notion in his chapter with reference to dissuading terrorist behavior.[29] And the use of military force does not necessarily signify a failure of coercion but is rather used as an instrument to underpin, establish, maintain, strengthen, and communicate *de facto* norms of behavior vis-à-vis violent non-state adversaries. By relying on the criminology literature in exploring contemporary Israeli deterrence of non-state militant groups, Thomas Rid has been particularly effective at squaring these conceptual circles.[30] Under certain conditions, punitive cumulative deterrence seems to make sense.

Flipping the discussion around, how might cumulative denial work? Most of the scholarship on cumulative coercion is based on the use of punishment and retaliation. Theoretically, however, threats of denial might also provide cumulative results, but they would rely on defensive rather than offensive action. And, as noted, half of Uri Bar-Joseph's original conception involves deterrence by denial at the strategic level: in the long term, military victory is deemed impossible, strategic failure all but assured. This is denial of long-term strategic and political objectives. Arguably, there may be room to expand cumulative deterrence to incorporate other aspects of denial more fully at the tactical and operational level. For example, tactical defensive capabilities that help shape behavior in the near-term might achieve coercive goals. Whereas cumulative deterrence by punishment is primarily based on promises of repeated pain, cumulative tactical denial would stem from the promise of repeated failure. And it would focus less on the horizon and more on denying immediate, on the ground, tactical objectives. Small and incremental defensive successes, say in repeatedly denying adversaries access to particular targets or denying them the specific tools or materiel they may need to achieve their goals, might add up, accumulate over time, and change behavior. The cumulative result would be deterrence by denial.

For illustration, let us return to the study of counterterrorism. Deterring terrorism by denial is often deemed a more reliable coercive strategy than relying on threats of punishment.[31] In terms of tactical denial, taking defensive measures to block terrorists (easy) access to their preferred targets tightens the security environment in which they operate and constrains the terrorism process as a result. Certain actions, like casing a target, recruiting and training operatives, and testing operations, may become increasingly hazardous. And certain types of attacks and those directed against certain targets become increasingly likely to fail. Militants are then forced to expend more time and energy preparing attacks that nonetheless appear likely to fail. Here again, Sawyer in his chapter explores these developments with reference to Israel and Northern Ireland.[32] In theory, the better a target is protected, the more complex attacks directed against it become, and the higher the level of operational risk (and failure). At some point militants will be forced to reassess the benefits of certain actions and/or attacks and may contemplate alternative behavior. Over time, these tactical defensive successes can accumulate, such that lasting changes in militant target selection and behavior becomes evident. Importantly, however, deterrence by denial in this particular case might otherwise lead to target displacement; that is, militants are deterred from some actions but go on to focus their efforts elsewhere. Other targets are attacked. So while security forces may have achieved a limited, cumulative, deterrence-by-denial success (i.e., terrorists elect not to attack a specific target or an entire class of targets), they have not deterred terrorists from conducting attacks writ large, or from developing more sophisticated attacks against hardened targets, or from attacking under-defended elements within the hardened target.

The denial–displacement paradox is both theoretically and practically pertinent. Theoretically it touches on the way scholars distinguish and explore deterrence-by-denial success and failure in counterterrorism: can deterrence succeed and fail at the same time? If, for instance, militants attack the Manhattan Mall after being deterred by heavy police presence from attacking a nearby subway station, would that count as a deterrence success? And practically speaking, there are limits to how far deterrence by denial and defense more generally can be applied. Limited resources are a fiscal reality. Not only are defenses spread thin, but Western states are target-rich environments.[33] Scholars and practitioners will have to discuss these points more thoroughly.

As an illustration of cumulative denial (and displacement), one Dima Adamsky and John Sawyer touch upon in both their chapters in this volume, consider that during the *al Aqsa Intifada* (2000–2005) Hamas, Palestinian Islamic Jihad, the al-Aqsa Martyrs Brigade, and other militant groups began by first attacking Israelis within Israel with suicide bombers dispatched against soft targets. Transportation hubs, bars, restaurants, and markets were repeatedly targeted in the first half of the conflict. In response, Israel brought to bear lessons it had culled from countering suicide bombers in Lebanon during the 1980s. During Israel's occupation of southern Lebanon, the IDF established denial and defensive mechanisms in theatre that were eventually instrumental in forcing Hezbollah to reassess the value and utility of deploying suicide bombers against IDF positions and personnel. Ariel Merari, in his 2000 testimony before the US House of Representatives, argued that suicide attacks dropped "very significantly" after Israeli measures "proved effective in preventing most of the suicide attacks." Hezbollah and others ceased using the tactic because they were "not bringing any results."[34] Fifteen years later, Israel would do much the same on its domestic front in dealing with an onslaught of Palestinian suicide bombers (over 110 such attacks were carried out in Israel and in Gaza/West Bank between 2000 and 2003).

Israel began by better defending public spaces: access to public transportation, universities, hospitals, and other such venues were restricted; guards were placed outside restaurants, bars, and markets; buildings were blast-proofed; checkpoints and security barriers were established between and within major cities; and privately owned shuttles (e.g., *Sherut*, a minivan, taxi-sharing service) were offered as alternative public transportation. The cumulative effect was the eventual reduction in the number of accessible soft targets. One tends to think that a guard outside a restaurant serves the diners inside: they are protected from harm because the guard is likely to stop a would-be bomber at the door (and suffer the consequences). But there is more to it: defenses go beyond simply protecting the target to manipulating an adversary's willingness to attack that target altogether. The first process is defense; the second is coercion. In Israel, over time, suicide bombers were forced to target military and police checkpoints over civilian targets. Other detonations occurred outside and off target. Some operations failed altogether, ending in capture. The cumulative result was a diminishment in the utility of carrying out suicide operations in Israel. Indeed, it has been years since Palestinian militants have carried out a suicide bombing in Israel. The benefit of conducting such attacks is uncertain and the risk of failure remains high. And yet, Israeli denial has altered but a tiny slice of militant behavior. Today rockets have replaced suicide bombings as the militants' preferred choice of weapons and tactic. This is the denial–displacement paradox put into practice. As Israel denied bombers access to targets, militants shifted their behavior to circumvent defenses. Rockets, while crude and less effective than suicide operatives, have proved effective instruments of terror.

COMMUNICATIVE DENIAL

Deterrence is not something that just happens. Rather, it needs to be actively applied, by one actor against another, and properly communicated. This is not to suggest that at times deterrence does indeed occur if left to its own devices. If a would-be victim (a defender) has "palpable strength," Freedman suggests, he may not have to do much to convince others to avoid attacking him. Here, deterrence is at work even without the defender making a clear point of communicating his strengths, capabilities, and demands. Freedman calls this "deterrence lite"; others might see parallels with general deterrence.[35] The deterrence puzzles that interest us, however, are usually those that are acutely participatory, where challengers issue threats and contemplate attacks and defenders mount counterthreats to dissuade them. In these cases, communication is a prerequisite to deterrence by punishment.[36] Challengers must appreciate what it is they are being deterred from doing (or are being compelled to do) and need to understand how action (or inaction in the case of compellence) will be met with retaliation. If threats and demands are miscommunicated, or fail to reach their intended audience, or are altogether un-communicated, deterrence in practice is likely to fail. In interstate deterrence, threats, demands, and capabilities are often communicated via policy declarations, and/or result from the illustrative signaling of military capabilities and intent. Signals can take various forms, including action—the mobilization or deployment of military assets; statements—public (and/or back-channel) announcements that outline or describe retaliation, ultimatums, and deadlines; and environmental conditions—both domestic and international support for the deterrent threat.[37]

What about deterrence by denial: how is it communicated? During the Cold War, part of NATO's conventional strategy involved denying the Soviet Army the ability to conduct a *blitzkrieg* into Western Europe. Soviet forces would have encountered a small but sophisticated NATO force capable of stalling and/or stopping a quick Soviet victory. The idea was to deny Moscow its preferred strategy.[38] NATO's denial (and punishment) capabilities, commitments, and credibility were communicated in a variety of ways: declarations of alliance were made, tested, and upheld; American troops and war materiel were stationed across Europe; and American-European economic integration was proposed and developed. Domestic defenses against more contemporary security threats—like mass-casualty terrorism—are similarly communicated. Verbally, US officials identify the steps they have taken to ensure terrorists are stopped at the border, that plots are uncovered, and that attacks fizzle. Furthermore, structural and mitigating defenses are visually apparent: cement and electronic bulwarks protect sensitive buildings (and even, entire cities) from attack; police detachments circle public areas; and first responders enter the fray immediately.

In discussing communicative denial, one issue worth exploring in greater detail is the question of palpability. Going back to Freedman, can denial be inherently "palpable"? That is, could deterrence by denial work without communication? Adversaries that remain unaware of the risks they take may be undeterred. As noted, when practicing deterrence by punishment, threats unmade will usually fail to alter a challenger's perception of cost. The same logic works with deterrence by denial: defensive capabilities that go uncommunicated will fail to alter a challenger's perception of probable benefits. The point of deterrence by denial is to manipulate an adversary's decision to try, not to leave them unaware of the defences they likely face, so that they try and then fail. But perhaps, as in the case of offensive military capabilities, communication and awareness is a matter of time and demonstration. After all, nobody questions that flint bows will fail to penetrate the hull of armoured vehicles. Here the denial capability is palpable, understood, inherent. It is rather the novel and adaptive defensive capabilities that need to be communicated, at least at first, so that potential adversaries understand evolving capacities and take them into consideration. Eventually these defences may be taken as a given and become part of the overall deterrence background.

Israel's Iron Dome is a case in point. Missile and rocket defense is nothing new. Israel, alongside its allies, has been working on the capability for decades.[39] The Iron Dome system targets and destroys short-range rockets, like katyushas, qassams, and other grad rockets, fired by militants. After years of tests the program quietly went operational in 2011. By the end of 2012, it had scored hundreds of intercepts in live combat operations. During Operation Pillar of Defense, the short November 2012 skirmish with Hamas, the Israeli Air Force reported that Iron Dome destroyed well over 400 rockets and missiles that would have otherwise landed in urban areas.[40] The defensive shield robbed Hamas of a strategic weapon and gave Israel some tactical breathing room. Unlike Israel's previous rocket wars and battles with Hamas (May and October 2004, June and November 2006, 2008–2009), Iron Dome gave Israel a chance to settle the conflict on favorable terms without having to resort to (re)invading parts of the Gaza Strip with ground troops. Media coverage of Israel's new defensive capability was widespread, domestically and internationally. Gob-smacked Western reporters in helmets and flak jackets craned their necks skyward as rockets were intercepted on live television. And hundreds of amateur videos were uploaded on YouTube.

Iron Dome was later deployed to Israel's borders with Egypt, Lebanon, and Syria, and to Jerusalem, Eilat, and Tel Aviv, and new batteries are in production. The message is that the technology works and that Israel is banking its future on it. Other complimentary anti-missile programs that provide Israel with layers of missile defense are also quietly being built and tested. For instance, David's Sling, Israel's response to medium-range cruise missiles, scored its first hit during a 2012 test; the Defense Ministry could not but help promote and disseminate a detailed video of the test.[41] Until Operation Pillar of Defense, Israel's missile defense capabilities were theoretical and not widely or well understood. Today, they represent—at least in appearance—a formidable technological impediment to militants, denying them the means to easily inflict harm on Israelis. In time, the demonstrative effect of Iron Dome and other missile defense programs may signal other groups, notably Hezbollah and Iran, that Israel has the ability to blunt the effects of their own missiles.

A separate communicative issue worth exploring is that of the audience. Deterrence is usually practiced against a specific challenger. Threats of punishment are tailored to the institutional and/or personal particularities of an actor's decision-making apparatus or leadership. Threats of denial, conversely, are increasingly applied against an entire class of actors and/ or actions. This is especially the case in deterring non-state and cyber-based groups. Threats of punishment target al Qaeda; threats of denial target terrorism. They speak to the same phenomenon but focus on different elements within it. The difference is subtle but important because it creates diverging priorities, results in different coercive behavior, and provides distinct challenges. In denying terrorism, the objective is to paint with broad strokes, to deny all actors intent on carrying out specific acts with the means to achieve their tactical and strategic goals. Herein, the deterrent message is that would-be targets are impenetrable, planned attacks will be botched, and attacks will fail. The audience is often unknown and includes any person or group intent on carrying out such attacks. To that end, denial mechanism can be broadly communicated to all challengers, equally and easily. Practicing deterrence by punishment, conversely, is a customized affair. Adversaries need to be identified and properly understood so that coercive threats can be specifically developed, credibly directed, and appropriately communicated. Punishment strategies require that states collect a great deal of information on the audience in question, on its objectives, goals, and structure. Threats will have to be precise, continuously updated, and repeatedly communicated. Moving forward in the study of deterrence, scholars may have to better address how the nature of the target audience, along with the type of threats employed and the domains in which these threats are deployed, affect the practice of punishment and denial.

CONCLUDING THOUGHTS

The study of contemporary deterrence by denial is in its infancy. The nature of contemporary threats along with the emerging security environment suggests a renewed impetus on the feasibility, practically, and utility of using threats of denial and failure to manipulate adversarial behaviour. This is not to suggest, however, that punishment strategies that rely on pain and retaliation have diminished in value. At times and under certain conditions punishment will continue to play an important role in deterring unwanted behavior. But elsewhere and under other conditions, denial may altogether trump punishment. And that may make the practice of deterrence, writ large, a more effective, robust, and prudent strategy. In many ways, then, denial is more advantageous than punishment.

Firstly, denial is closely related to defence. True, the two are conceptually distinct: defence mitigates the effects of an attack whereas denial influences an adversary's decision to launch an attack. Yet defense is also what happens when deterrence by denial fails and attacks are launched.[42] Herein, it acts as a security safety-net, limiting the consequences and costs of practicing deterrence in a way that deterrence by punishment cannot. Secondly, deterrence by denial may be the unintended and unplanned result of other, offensive- and defensive-minded security operations. For example, operations meant to disrupt terrorist networks—like arrests, counter-financing, and targeted killings—can have a denial effect even though deterrence is not the primary objective. The intention, in this case, is to round up operatives and starve the organization of cash and leadership. The effect, however, is to also simultaneously deny the group what it needs to implement its preferred strategy and achieve its tactical and strategic goals.[43] Here offense, defense, and denial intermingle in unique ways that not only impede violence outright but also potentially manipulate a group's willingness to use violence.

Thirdly, denial's ascent suggests there may be opportunities to better integrate it with deterrence by punishment, at a level playing field, that might strengthen the practice of deterrence overall. Each supports the other towards a shared and common coercive goal. This is not a particularly novel suggestion; Iraq's non-use of chemical weapons in 1990-1991, and NATO's strategy for protecting Europe against the Soviets, are cases in point. But in tackling contemporary threats it may be especially important for defenders to hedge their bets by practicing both together.[44] In counterterrorism and cybersecurity, for instance, denial alone may not be enough. Indeed, some terrorist groups may see value in conducting attacks that are nonetheless likely to fail because they equate the sensationalism of the attack as a value in and of itself. Al Qaeda, IS, and other groups, for example, repeatedly target commercial aircraft with new devices despite a losing track record stretching over a decade because the plots themselves generate value. In the cyber realm, denial alone may work poorly in deterring challengers because the costs of launching attacks, as Libicki reminds us in his contribution, are exceptionally low and are traditionally, risk-free.[45] One way to balance the scale is to add punishment to the equation, to promise aggressors both defeat and retaliation simultaneously. In sum, that deterrence by denial in both theory and practice shows signs of renewal for countering emerging security threats will benefit the practice of deterrence in international relations writ large.

Notes

1. Zagare, "Deterrence is Dead. Long Live Deterrence"; Wilner, "Deterring the Undeterrable."
2. Chalmers, "Deterrence and Counterproliferation," 149. See chapters 1, 5, and 6.
3. See chapter 1.
4. See chapters 6, 7, and 8.
5. Freedman, *Deterrence*, chapter 2; Terrill, "Escalation and Intrawar Deterrence during Limited Wars in the Middle East." For details on intrawar deterrence by punishment applied to counterterrorism, see Wilner, "Fencing in Warfare." Brief parts of the following discussion were first explored in the article.
6. Chilton and Weaver "Waging Deterrence in the Twenty-first Century," 35.
7. Freedman, *Deterrence*, 32–34. Others have used the narrow–broad dichotomy to describe other deterrent functions, confusing the typology somewhat. Thomas, Kiser, and Casebeer, *Warlord Rising*, 186–188.
8. Snyder, *Soviet Strategic Culture*, v.
9. Kahn, *On Escalation*.
10. Posen, "US Security Policy in a Nuclear-Armed World," 175. An earlier version of Posen's chapter appeared as "U.S. security policy in a nuclear-armed world or: What if Iraq had had nuclear weapons?" *Security Studies* 6:3 (1997).
11. Freedman, *Deterrence*; Posen, "U.S. security policy"; Davis and Arquilla, *Deterring or Coercing Opponents in Crisis*.
12. Chemical agents were used, most notably, during Fascist Italy's invasion of Abyssinia (Ethiopia) (1935–1936), Imperial Japan's war with China (1937–1941), Egypt's involvement in the Yemen Civil War (1963–1967), and Iraq's attacks on Iran during the Iraq–Iran War (1980–88) and against the Kurdish village of Halabja (1988). Syria used chemical agents throughout its recent and ongoing civil war (2011–). And terrorist groups, too, have conducted chemical attacks: Aum Shinrikyo's sarin gas attack in Tokyo, Japan (1994) and Al Qaeda in Iraq's widespread use of chlorine bombs (2006–2007) stand out.
13. Schneider, "Deterrence and Saddam Hussein," 29–30, note 53.

14. For a compelling look at U.S. efforts to deter—and later, compel—Iraq before and during the Gulf War, see Gross Stein "Deterrence and Compellence in the Gulf, 1990-91."
15. President George H. W. Bush, "A Warning Letter to Saddam Hussein," in Schnieder, "Deterrence," Appendix B.
16. Baker, *The Politics of Diplomacy*, 437.
17. Toth, "American Support Grows for Use of Nuclear Arms."
18. Tellingly, however, two of President Bush's other January 5th threats—that the U.S. would not tolerate "the destruction of Kuwait's oil fields" and that Hussein "would be held directly responsible for terrorist actions against any member of the coalition"—went unheeded. Retreating Iraqi forces did set fire to Kuwait's oil infrastructure, to devastating effect. And Hussein, through the Iraqi Intelligence Service (IIS), plotted to assassinate (former) President Bush during his April 1993 visit to Kuwait to commemorate the Coalitions' victory over Iraq. The plan—to detonate a car bomb in President Bush's vicinity—was uncovered. The US, under President Bill Clinton, retaliated by launching 23 cruise missiles against ISS facilities in Baghdad on June 26, 1993.
19. Schnieder, "Deterrence and Saddam Hussein," 33; Mauroni, *Chemical-Biological Defense.*
20. Schneider, "Deterrence and Saddam Hussein," 28
21. Other explanations abound, of course. Perhaps Iraq was retaining its CW to deter the U.S. from moving into Iraq (and removing Saddam Hussein) once Kuwait had been liberated. Or perhaps the conditions were poor for CW use—Jonathan Tucker explains that the prevailing winds which had blown "out of Iraq" during the crisis shifted to the southeast towards Iraqi lines at the onset of hostilities. Tucker, "Evidence Iraq Used Chemical Weapons," 114.
22. The literature on the subject of reputation is vast. For an early debate, see the special issue of *Security Studies* 7:1 (1997). For a more recent discussion, see Sartori, *Deterrence by Diplomacy.* Patrick Morgan, too, wrote about iterated, low-level engagements as they relate to enforcing threats and communicating deterrence. See Morgan, *Deterrence Now*, 265–268.
23. See chapter 7.
24. Bar-Joseph, "Variations on a Theme," 156–157.
25. Almog, "Cumulative Deterrence and the War on Terrorism"; Bar, "Deterrence of Palestinian Terrorism"; Davis. "Towards an Analytical Basis for Influence Strategy in Counterterrorism," 85; Rid, "Deterrence Beyond the State," 128–130, 137–142

26. Crenshaw, "Will Threats Deter Nuclear Terrorism?", 143.
27. Gross Stein, "Deterring Terrorism, Not Terrorists."
28. Knopf has written about this in several places; see "The Fourth Wave in Deterrence Research" and "Terrorism and the Fourth Wave in Deterrence Research."
29. See chapter 4.
30. Rid, "Deterrence Beyond the State," 125.
31. Numerous explanations have been proposed: terrorism is often coordinated anonymously which makes attribution and punishment difficult; terrorists often lack a "return address" for retaliation; terrorists have "enormous counter-coercion potential" which undermine certain retaliatory strategies; militant leaders may be willing "to lose it all" to conduct spectacular attacks; militants may actually want retaliation to galvanize supporters; terrorism may be used for internal, group-based dynamics, rather than to achieve strategic purposes; and policies that seek to defeat terrorists undermine the logic of coercion, which requires inaction in the event demands are met. See Kapur, "Deterring Nuclear Terrorists"; Morral and Jackson, "Understanding the Role of Deterrence," 7–8, 16; Smith, "Strategic Analysis, WMD Terrorism" 159–179; Smith and Talbot, "Terrorism and Deterrence by Denial," 54–59; Auerswald, "Deterring Nonstate WMD Attacks," 545; Trager and Zagorcheva, "Deterring Terrorism," 92–93, 108–111; Wehling, "A Toxic Cloud of Mystery, 288–291; National War College, *Combating Terrorism in a Globalized World*, 42; Morral and Jackson, "Understanding the Role of Deterrence," 7; Betts, "The Soft Underbelly of American Primacy," 19–36; Schaub, "When is Deterrence Necessary," 64–69; Frank Harvey and Alex Wilner "Counter-Coercion, the Power of Failure," 95–111; Wilner, "Fencing in Warfare"; Wilner, *Deterring Rational Fanatics*, chapters 2 and 3.
32. See chapter 4.
33. See Betts' discussion of power asymmetry and offense-defense in counterterrorism in Betts, "The Soft Underbelly of American Primacy," 26–33.
34. Merari, Testimony before the Special Oversight Panel on Terrorism, U.S. House of Representative, July 13, 2000.
35. Freedman, "Framing Strategic Deterrence," 47.
36. See my discussion of this with Frank Harvey in "Counter-Coercion."
37. Harvey, "Rigor Mortis or Rigor," 676.
38. There was more to the strategy, of course. Denial was interlinked with the threat of nuclear escalation and retaliation: The Soviets risked starting a general war that would ultimately force NATO to use nuclear

weapons down the line. Mearsheimer, "Prospects for Conventional Deterrence in Europe," 158; Gerson, "Conventional Deterrence in the Second Nuclear Age," 32–48; Huntington, "Conventional Deterrence and Conventional Retaliation," 35–40; Harknett, "The Logic of Conventional Deterrence."

39. See Adamsky's chapter on the Israeli program, and Trexel's chapter on the Japanese (and American) programs for details, chapters 6 and 7.

40. IAF, "Events Log," http://www.iaf.org.il/4388-39969-en/IAF.aspx, accessed Aug 2019.

41. A video of the intercept is available at: *Times of Israel*, "David's Sling Success Caught on Film," (November 27, 2012).

42. Snyder, *Deterrence and Defense,* 3. See also Gerson, "Conventional Deterrence in the Second Nuclear Age," 38.

43. Keith Payne writes: "... denial frequently is the concomitant effect of military or law enforcement efforts aimed at defending against or defeating non-state actors." Payne, et al., "Deterrence and Coercion of Non-State Actors," 37.

44. Gerson, "Conventional Deterrence in the Second Nuclear Age," 37.

45. Glaser, "Deterrence of Cyber Attacks and U.S. National Security," 2. See chapter 8 of this volume.

CHAPTER 3

THE SOCIAL PSYCHOLOGY OF DENIAL

DETERRING TERRORISM[1]

Janice Gross Stein and Ron Levi

Every time passengers check in at an international airport almost anywhere in the world today, they directly experience a strategy of deterrence by denial. The removal of electronics from personal luggage, the shoes and belts in the plastic bins, the scanners and the body searchers are all designed to convince would-be attackers that their chances of successfully hijacking an aircraft are low. Denial strategies are different from strategies that seek to deter by punishment, which threaten that if a would-be attacker strikes, the costs that will be inflicted in reaction will far outweigh the benefits.

Strategies of deterrence by punishment are common in the vocabulary of security: if you do what I do not want you to do, then I will punish you so that the costs exceed any benefits that you anticipate from your action. Deterrence by punishment is conditional: they are always "if

... then ... " statements. Strategies of denial work differently; they are unconditional and always in place. Airport security does not diminish perceptibly even when there is no evidence of an imminent attack. On the contrary, deterrence by denial works because a would-be attacker always estimates the probability of failure as low. The benefits of hijacking or exploding a passenger jet are not less now than they were a decade ago; rather, the probability of failure is far higher because of ubiquitous and thorough inspections. The manipulation of estimates of probability is doing most of the theoretical work in deterrence by denial.

In an era of transnational terrorism and frequent cyberattacks against civilian as well as military infrastructure, deterrence by denial has become newly prominent. The underlying theoretical logic of deterrence by denial is the same if the target is a state or a non-state actor, but its application differs. We understand terrorism as a strategy of political theater, inflicting punishment on innocent civilians, on bystanders who are not directly involved in a conflict, to delegitimize leaders or governments by alienating and frightening their populations.[2] It can best be understood as a process over time, as political strategy in asymmetrical conflict. We give special emphasis to delegitimation—and thereby destabilization—as a goal of transnational militants and shadowy hackers who engage in acts of terror against states. The struggle for legitimacy, we contend, becomes one of the critical theaters of contestation.

THE LOGIC OF DETERRENCE BY DENIAL

To deter terrorism, governments have used strategies of both punishment and denial. In this chapter, we focus on the theoretical arguments and practical applications of denial and broaden the construction of deterrence by denial to include deterrence of terrorism through delegitimation of terror as a political strategy.[3] In response to those who seek to destroy the legitimacy of governance by terrorizing populations, delegitimation in turn is increasingly important as one of the deterrence-by-denial strategies in governments' repertoires.

Governments have made massive investments in homeland security in the last decade. They have increased airport inspections, hardened the perimeters of critical infrastructure, invested heavily in intelligence collection and analysis, and built firewalls in cyberspace.[4] Interpreted through the lens of rational-actor theories, these tactics play principally on increasing the likelihood of failure, and only indirectly on cost, in seeking to reduce the likelihood of attack.[5]

What is required, in theory, for a strategy of deterrence by denial to succeed? Beyond the usual requirements of clear, credible, and costly signals, the risk calculus that is at the heart of deterrence by denial remains opaque. It seems reasonable to assume that adversaries must be motivated to attack; able to read the signs of added protection of targets, massive data collection and analysis, and strengthened firewalls; and, finally, able to calculate, with rough approximation, the probabilities of failure. But even this relatively relaxed formulation raises a host of theoretical questions.

Most importantly, what probability of failure becomes unacceptable? How do theoretical models specify these thresholds? What body of theoretical literature can we use to determine whether an al Qaeda cell considers an 80 percent likelihood of failure as too high, or as an acceptable risk? In rational choice specification of deterrence by punishment, the theoretical logic is clear: the subjective expected costs are greater than the expected benefits. In rational choice specification of deterrence by denial, does the probability of failure simply have to be greater than success? The literature is silent on risk propensity and its implications for choice.

Theoretically, highly motivated attackers may accept a very low probability of success since they can afford to fail many times so long as they succeed once. Even a high probability of failure may be acceptable because even when these attacks fail, they frighten the public, magnify the reputation of the attacking group, and generally contribute to the theatricality of terrorism that is such an important part of strategy.[6] In the first case, the focus is on the large benefits of success even when the probability of success is low, similar to a low-probability but high-cost event. We insure against these kinds of events all the time. In the second case, failure brings benefits as well as costs. We have no evidence to suggest whether and when these different kinds of calculation govern the choices of those orchestrating acts of terror.

One other argument has been leveled against the theory and strategy of deterrence by denial. No matter how impressive the defense, no matter how low the likelihood of success, highly motivated attackers probe for weakness and ultimately "design around" the defense. No defense is impregnable, as the history of the Maginot Line in France in 1940 and the Bar-Lev line in the Sinai Peninsula in 1973 attest. This is all the more so when highly motivated attackers can probe multiple points of access and find the weakest link in the chain. It is no surprise, for example, that audits of airport security systems routinely find holes in the system which make the system as a whole vulnerable.[7]

Scholars in international relations have also suggested that when the odds of failure become unacceptably high, would-be attackers displace their action to less-defended, softer targets.[8] When airport security becomes too difficult to penetrate, subway systems, crowded plazas, shopping malls, and sporting events, all largely undefended, become targets of opportunity. If correct, evidence of displacement would not challenge the theory of deterrence by denial, but it would significantly limit its benefits.

Analysis of deterrence by denial, particularly deterrence of attacks against civilians by militant organizations, suffers from the well-known challenge of identifying cases of "success." When deterrence succeeds, the dog does not bark. We have little or no access to evidence from al Qaeda or other organizations that suggests whether and when they have been deterred from attacking because of a perceived high likelihood of failure that is due directly to actions that a defender has taken to harden targets, improve intelligence, or otherwise secure vulnerable sites. We do have evidence of attacks that were not completed, whether they were thwarted at the planning stage or known to have been attempted and failed. In 2006, al Qaeda militants tried to detonate liquid bombs on transatlantic aircraft.[9] They tried again in 2009 with the "underwear bomber," and yet again in 2010 with explosive printer cartridges shipped on cargo planes.[10] That these attacks were not successful speaks to the success of intelligence agencies in collecting and analyzing data and in penetrating these networks, as well as to plain dumb luck. However, these attempts also speak to the limits of deterrence by denial to deter challengers from attempting to hijack aircraft.

EXAMINING DETERRENCE BY DENIAL THROUGH THE LENS OF CRIMINOLOGY

If aborted attacks testify to the failure of deterrence, we cannot assume by default that long periods of quiescence or inactivity establish the success of deterrence. Identifying cases of successful deterrence over time has long plagued the analysis of deterrence. We propose to come at the analysis of deterrence of terrorism by denial from a different perspective. Some scholars working on terrorism argue that terrorism is best analyzed within the framework of criminal activity and, correspondingly, that terrorist actions are best dealt with through the criminal justice system rather than through extrajudicial mechanisms.[11] Whatever the merits of the strategic arguments, the theoretical argument for deterrence strategies derived from the criminal justice paradigm is intriguing and compelling. Indeed, criminologists have begun to assess counterterrorism policy within the context of other crime prevention strategies, although they do not relate their analyses to the ongoing debates in international relations.[12] Locating terrorism as a subset of criminal activity allows for comparison of the effectiveness of deterrence by denial across similar types with robust empirical studies. Although there are, of course, concerns over whether research insights from studies of one type of crime are applicable to other forms of criminal activity, these challenges are arguably less significant than relying exclusively on logical argument or on the limited and anecdotal data we currently have on the effectiveness of deterrence by denial against global terrorism. As we marshal data from studies in criminology, we also examine the complementarity and overlap of theoretical arguments from criminology with those from the deterrence literature in international politics.

In separate work, we examine the arguments relating to displacement, which are common to the literature in both international politics and criminology.[13] Drawing on studies of displacement and diffusion in criminology, we find that the conventional wisdom about displacement in the deterrence literature in international politics may well be wrong. Instead, studies suggest the intriguing and counterintuitive argument that focused crime prevention, or deterrence by denial, may diffuse the benefits of crime prevention to nearby areas.[14] Experimental and quasi-experimental research in the United States and the United Kingdom consistently demonstrate that geographically based policing interventions—for which one can measure the effect on crime both locally and beyond the geographical scope of the intervention—does not displace crime between locations.[15] If anything, the strongest evidence from randomized control trials points to a diffusion of crime control benefits,[16] so that these interventions lead to a drop in crime in *both* targeted and nearby locales.[17]

We also looked at evidence drawn from experimental studies that speak directly or indirectly to strategies designed to increase the likelihood of failure by deliberate manipulation of the physical and social environments to deny would-be offenders the opportunity to commit criminal acts. We paid particular attention to evidence from criminology on public surveillance, especially on the use of CCTV cameras and whether, when, and why they work. The arguments and evidence from criminology speak to the strategy of "target-hardening," one of the key strategies in the literature on deterrence of terrorism through denial, which is designed to increase the likelihood of failure. We found some evidence of success but far less than is generally assumed.[18] Quasi-experimental studies in

Canada, the United Kingdom, and the United States all demonstrate that CCTV is not particularly effective in reducing crime, except for the limited case of CCTV in reducing auto thefts on streets and in car parks.[19]

Deterrence-by-denial strategies are thus broadly supported by this experimental and quasi-experimental work in criminology. First, the possibility of crime displacement—a potential, unintended, and negative consequence of any successful denial strategy—seems unlikely; if anything, criminological evidence points to a potential diffusion of crime control benefits. Second, public surveillance through CCTV tends to be effective only in those cases where offenders face a higher probability of immediate failure, such as when curtailing automobile theft through enhanced lighting strategies in parking lots, but is largely ineffective when it merely increases the likelihood of detection and eventual capture after the offense is committed. And third, target-hardening is most successful when it involves physical denial strategies, with largely successful, albeit mixed, results in cases of human guardianship and monitoring (whether these are formal guardians, such as security guards, or the presence of adult residents on city streets).

In this chapter, we extend our focus to examine studies of deterrence by denial that both increase the risks of failure and reduce the social rewards to perpetrators of acts of terror. Scholars of deterrence have only recently begun to develop theoretical arguments about deterring terror by delegitimation. The purpose is to manipulate the social rewards of prohibited action: "The objective is to reduce the challenger's probability of achieving his goals by attacking the legitimacy of the beliefs that inform his behavior."[20] We focus on three approaches that all work to deter, not only by influencing estimated probabilities of failure through the hardening of targets and surveillance—as the established theoretical literature in international relations does—but also by influencing individual and collective beliefs about social risks and reward. The factors influencing these beliefs are individual belief in the legitimacy of law enforcement, social bonds and informal controls that are activated by people who are influential in the community within neighborhoods, and ecological outcomes that attend living in neighborhoods rich in collective efficacy[21] with high levels of social cohesion.[22] Criminological research has determined that at each of these levels, deterrence by denial can be enhanced by acting directly or indirectly on the social risks and rewards of would-be perpetrators.

We begin by examining studies of the impact of procedural justice on preventing crime in neighborhoods. Studies from criminology speak to the willingness of community members to cooperate with law enforcement and to share information with the authorities. This kind of information is becoming increasingly important as young recruits stream into conflict theatres like Syria and then return with new organizational and leadership skills that may result in greater security challenges. These studies speak to deterrence by denial by increasing the likelihood of failure.

We then turn to the effectiveness of reducing the social rewards of acts of terror by activating existing social bonds within communities and changing the available "norms and narratives" on which community members (and would-be offenders) draw in making choices about appropriate behavior. We examine the deterrent effect of having community members articulate norms, expectations, and informal sanctions for criminal acts. Closely related are arguments in criminology on strategies designed to induce shame. There is a growing literature on rule-setting and strengthening social condemnation of violent and criminal acts to remove social supports from those who commit criminal acts. The evidence here is less robust but nevertheless suggestive of the capacity for engaging a deterrence-by-denial strategy through delegitimation within communities.

We move next to the ecological level of the community, where criminologists find that neighborhoods rich in collective efficacy enjoy salutary effects on crime and produce environments that reduce legal cynicism among residents. Robust findings in criminology demonstrate that informal social control is not only a characteristic of individual beliefs or social bonds but also "rooted in shared expectations and perceived codes of conduct."[23] Deterrence by denial is made possible by activating and making salient a shared moral context in the community where people reside, thereby reducing the social rewards of criminal action.

Finally, as a contrast to these three approaches, we examine evidence that changes in policing presence and perceptions of public order and disorder have an impact on the prevention of criminal activity. Here, we draw on work that assesses the evidence on "broken windows" policing and the effectiveness of strategies to police disorder. Research finds these "broken windows" strategies to be ineffective or at best inconclusive in reducing crime. The assertion of visible symbols of protection through deployment of police officers is not enough, and requires, in addition, attending to processes of social reward and sanction.

Formal Order Maintenance: The Role of Procedural Justice

The degree to which community members cooperate with law enforcement is central to the capacity to deter by denial. Community cooperation in the deterrence of terror by denial is important in two distinct ways. First, when community members engage effectively with police, they can mobilize to warn of suspicious packages, compromised security, or other indications that targets are vulnerable. In Israel, members of the public routinely warn law enforcement when they see unattended packages aboard public transport or in public places.[24] Community members must feel a sense of efficacy and responsibility if they are to engage in this kind of activity.

Even more important, community members can provide early warning of young people from their own community who are likely to be involved in acts of terror. Deterrence by denial works here by increasing police capacity to obtain information to thwart would-be terrorists, rather than by changing their risk calculus. Scholars emphasize the importance to counterterrorism of police being able to elicit information and internal cooperation from within communities.[25]

Engagement with law enforcement is often undercut by widespread cynicism within communities over the role of the police.[26] There are two strands of research in criminology that are relevant: the first, at the individual level, demonstrates the importance of beliefs about procedural justice, and the second, at the ecological level, demonstrates the importance of community norms of collective efficacy. Both provide empirical evidence about the mechanisms through which community members either cooperate with the police or engage in the informal maintenance of order, even in highly challenging social and economic contexts.

A rich vein of research consistently demonstrates the importance of procedural justice for individuals' willingness to cooperate with the police and to comply with legal rules.[27] Procedural justice rests on whether people judge police processes and law enforcement as fair, even if they disagree with the substantive outcome. It draws on people's beliefs about neutrality, their respect, and their trust. These beliefs and attitudes are more important for legitimacy—or the acceptance of decision-making —than their views of the substantive outcome or even their stake in winning or losing. This emphasis on beliefs about the process, rather than on the attractiveness of the outcome, remains the case even when the stakes are high in cases of incarceration and high financial stakes, and when important policy issues are being decided.[28] Perceptions of procedural justice help explain why people voluntarily obey the law and regard state power as legitimate.[29]

As a cornerstone of legitimacy, perceptions of procedural justice are also critical to explaining why people cooperate with the police in responding to crime.[30] A two-wave panel study drawn from a random sample of New York City residents asked about their willingness to cooperate with the police by calling the police to report that a crime was occurring, helping the police to find a criminal, reporting suspicious activity to the police, volunteering time to help the police, patrolling the streets with others, and attending community police meetings about crime.[31] Residents who

in the first wave of interviews regarded the police as more legitimate—
that is, that they make decisions fairly and treat people justly—were
found in the second wave of interviews, one year later, to be more likely
to cooperate with the police.[32] These models rely on people's beliefs
about what is appropriate, rather than on instrumental calculations.

Positive perceptions of procedural justice are a precondition for engagement with the police and willingness to supply information in counterterrorism investigations. Criminologists have directly investigated
whether minority groups will comply with the police in the context of
counterterrorism investigations. A series of studies that include Muslim
communities in both London and New York find that, for Muslim Americans, it is the belief that police processes are fair and that people are
treated fairly during the decision-making process that induces cooperation, either as a direct consequence of this fairness, or because this
perception of fairness leads to a more general belief in the legitimacy
of the police themselves.[33] Belief in police legitimacy changes behavior;
it induces people to defer to authorities and to cooperate voluntarily
with the police.[34] These studies demonstrate that the political ideology
of respondents—their views on foreign policy, their attitudes toward
terrorism, and their religious identity—generally has no effect on willingness to engage with law enforcement in the United Kingdom and has
only limited impact in the United States.[35] When people do not perceive
police intervention as harassment, their membership in a group that
is targeted for increased police attention appears not to change their
assessment of procedural justice.[36]

In contrast, these same studies demonstrate that instrumental mechanisms—whether residents believe that police were effective, or even
whether they believe that terrorism is a serious problem—are only weakly
and indirectly related to cooperation. For both majority and minority
group members, when people believe that Muslim Americans are being
unfairly targeted or harassed, there are negative spillovers for police
legitimacy. Beliefs about the consistent elements of procedural justice—

neutrality, respect, and trust—remain central in fostering cooperation for counterterrorism as well as crime control. Neither the degree of violence in acts of terrorism nor the identity of the respondents as Muslim Americans changes the central importance of procedural justice for predicting cooperation with the police.[37]

One way deterrence by denial works is by building strategic contacts between police and community members and fostering public trust of the police within communities whose members may be reserved about, or even fearful of, engagement with law enforcement officers. Drawing on empirical evidence from research on procedural justice, we identify the creation of "community intelligence" as one in the broader menu of strategies of deterrence by denial.[38] Community members in the United States and Great Britain who consider the police fair are more likely both to alert the police and to cooperate with law enforcement about potential security threats.[39] The evidence suggests that the success of deterrence by denial rests not only on a credible threat of failure but also on strongly embedded norms of police fairness within communities.

Community Norms and Shaming

A different strand of research focuses on inducing shame in a would-be attacker for the violation of community norms.[40] This strategy denies opportunity by delegitimating acts of terror and increasing the likelihood of social sanctions against those who would commit acts that are normatively impermissible. The probability of failure increases not by what those who seek to defend against terror do but rather by what communities with high levels of social cohesion do. Scholars emphasize that marginalizing terrorist groups from their imagined constituencies is central to counterterrorism efforts.[41] This kind of strategy has been described as deterrence by delegitimation; we examine its likely effectiveness as a form of social-psychological denial that stems from the community of origin of individuals who contemplate acts of terror.

How does a strategy of shaming work? Community leaders engage in rule-setting or reducing uncertainty about the impermissibility of certain acts. This kind of rule-setting has been of increasing importance in the global Muslim community, where religious leaders have explained that killing innocent civilians is against Islamic law and have issued *fatwas* against these kinds of acts.[42] Rule-setting of this kind is especially important when there is uncertainty about what is permissible. Many young Muslim men who have been mobilized to commit acts of violence have been told by organizers that their acts of violence are divinely sanctioned and a fulfillment of Islamic law.[43] Clarifying religious rules can be effective for young men motivated in part by religious beliefs.

How best to enable the development of shared norms that strengthen the condemnation of violent acts and the removal of social supports? The evidence in criminology is thin on this point, but research suggests that neighborhoods rich in shared norms of social order and of supervision of young people are highly correlated with communities with high levels of intergenerational closure, where local adults know and are a resource to support children and adolescents in the community, and reciprocated exchange, where interfamily interactions can build parenting supports across families.[44] In these kinds of contexts, local adults are likely to be more successful in building norms of social order and supervision.

Closely related to developing norms of social order is strengthening the social condemnation of impermissible acts. It is not only that these acts are unlawful but also that they violate community norms and values. Here, communities make the foundations of their moral reasoning explicit to establish a set of shared normative constraints on what they consider impermissible behavior. Criminological research that focuses on the impact of "pulling available levers"—repressive policing combined with positive opportunities—to combat open-air drug markets and crime-ridden gang areas suggests that this kind of strategy is promising.[45] These strategies rely both on formal policing efforts, job opportunities, and training, and on a "norms and narratives" approach that mobilizes community-based influentials to denounce offenders and to signal the community's disapproval of the behavior.[46] Community influentials, such as family members, faith leaders, ex-offenders, loved ones, and others, speak to offenders directly to deliver messages such as, "We love and care about you. We want you to succeed. We need you alive and out of jail. But if you do not absolutely understand that we disapprove of what you are doing, we are going to set that straight today."[47]

By emphasizing "norms and narratives," these crime prevention strategies work at one level to activate social bonds in the community, a strategy that has a long history in criminological research on informal social control. Community influentials promote attachments to families, commitments to social relationships in communities, involvement in legitimate employment, reinforcement of social norms, and prosocial attitudes based on a belief in law-abiding behavior.[48] Yet this strategy goes further by focusing on the content of the messages being conveyed by community influentials: these offenders are indeed valuable to the community and to those who enforce the law (despite the prevailing narratives that structure the relationships between community residents and law enforcement), but the community also "needs the violence to stop," "the ideas of the street code are wrong,"[49] and the community influential is hopeful about the offender's potential future. A "norms and narratives" strategy mobilizes the moral voice of the community to denounce offenses and condemn violence, and to counter prevailing local narratives of, for example, the acceptability of drug dealing or prevailing local norms of gang retaliation. This is a strategy that, at the informal level, mirrors deterrence-by-denial strategies by removing social approval. Indeed, this kind of strategy goes beyond constraining likely individual offenders, since, by removing the social reward, it has the potential to reduce the likelihood of imitation.

A central premise of this approach is that gang and drug crime is closely linked to informal codes that permit and reward violent criminal behavior. These street codes emphasize honor and retaliation, community anger, frustration, and suspicion of law enforcement that is perceived as racially biased and heavy-handed, as well as some tolerance within the community for serving prison time. This strategy gives greater weight to codes of honor, status, and informal solidarities, rather than drug markets as contributors to violence. Influential community members, including faith leaders, are mobilized in these programs to draw out and express underlying positive community voices and expectations, as are ex-offenders and other community and family members. They are mobilized along three dimensions: they set the community standard by indicating, "We need you, and you're better than this"; they provide moral engagement by invoking the possibility of offenders' mothers or community children being put in harm's way; and they challenge street codes by asking offenders who will help their families when they are imprisoned and condemning their gang violence as politically unjustifiable.[50]

In the words of David Kennedy, the leading designer and proponent of these programs, "You could *feel* the room change ... We started calling it *the moral voice of the community*. Moral engagement with thugs. The very last thing anybody thought you could do."[51]

Because these strategies rely on several "levers" to address crime and violence—from policing and enforcement actions to job training to community-based disapproval—it is methodologically challenging to identify empirically the successful elements of these programs. Program proponents point to reductions in drug-related and violent crime of between 40 and 50 percent, but they do not tease out the elements of the program that are key to its success.[52] Program evaluations by other researchers carefully delineate more modest but substantial program successes of approximately 12 percent, though even then it is challenging to distinguish which elements of the program account for the impact.[53] As one analyst of these programs suggests, "In the focused deterrence approach, the emphasis is not only on increasing the risk of offending but also on decreasing opportunity structures for violence, deflecting offenders away from crime, increasing the collective efficacy of communities, and increasing the legitimacy of police actions. Indeed, it seems likely that the observed crime control gains come precisely from the multifaceted ways in which this program influences criminals."[54]

Shaming, the articulation of prosocial norms and narratives, the assertion of personal and community bonds, and beliefs about procedural justice are joined up in these strategies of focused deterrence and pulling levers. Taken together, these strategies reflect social-psychological and normative approaches to deterrence by denial and delegitimation.

Collective Efficacy and Informal Social Control

In addition to the individual level of belief in legitimacy and trust in law enforcement and the bonds and norms promoted directly to would-be offenders by influential community members, research in criminology further points to the informal social controls that are rooted in the ecological level of neighborhood life. These social controls enable strategies of deterrence by denial and legitimation. "Thicker" strategies of problem-oriented policing that do not rely exclusively on surveillance or target-hardening but also include police involvement to help address community problems, appear more likely to achieve a broader geographic diffusion of benefits.[55] These programs include a reliance on community advice that gives police the information they need to develop a tailored response and monitor and assess the results of their strategy.[56] These approaches are rooted more deeply in the wider context of community engagement and the building of informal social controls.

A related series of studies in criminology demonstrates the centrality of collective efficacy on neighborhood outcomes, including reduced levels of violence. The study of neighborhood effects, including the ecological capacity of neighborhoods to engage in informal social control, finds that safety and security are a function of a neighborhood's collective efficacy.[57] Collective efficacy requires the activation of social ties and the generalized belief that people in the neighborhood will act on behalf of others and the collective. A neighborhood's collective efficacy is a measure of the social cohesion among residents, combined with shared expectations that residents will intervene on behalf of the common good to monitor children, respond to disorder, intervene to stop violence, or secure city services. Collective efficacy thus speaks not only to the building of trust and cohesion but also to moral context and the capacity for deterrence. As Robert Sampson, the leading researcher in this field, recently explained, "A key argument of collective efficacy theory is that it matters what I think others think, making collective efficacy a kind of deterrence or moral rule—a generalized mechanism of 'common knowledge' that goes beyond any single act of control."[58] This is deterrence through denial and delegitimation.

Collective efficacy is a consistent and significant predictor of how neighborhoods successfully deter crime and disorder—including in neighborhoods of high and concentrated forms of disadvantage—by increasing informal social control.[59] Evidence from the United Kingdom suggests that such informal, neighborhood-driven collective efficacy also predicts public confidence and perceived legitimacy of the police.[60] And, recent evidence from the United States demonstrates that the lack of informal social controls is correlated at the micro-level, across 24,000 street segments over a sixteen-year period, with chronic crime hot spots. This evidence suggests that highly targeted social interventions to bolster informal social control may be a promising crime prevention strategy.[61] This vein of research in criminology is robust and promising for scholars of deterrence by denial. Research identifies community collaboration as essential to engagement with the police and other state officials trying to deter violent acts of terror by changing collective beliefs and thereby raising the likelihood and the costs of failure. Studies demonstrate convincingly that a perception of procedural justice is necessary for individual engagement. At the community level, a sense of collective efficacy is essential, as is a shared belief in the legitimacy of law and a sense of shared norms.

How is a sense of collective efficacy enabled? Several strands of research are suggestive here. Criminologists suggest that the most effective strategies are likely to be those that help communities to help themselves.[62] Institutional integration between the police and community organizations can give new meaning to law enforcement and public safety interventions.[63] Rather than a public-centered notion of law enforcement, which envisions the police as the primary agents of social control through the use of a politically legitimized monopoly on force, cooperative alliances among community organizations that are facilitated by government can set the stage for "private" law enforcement, where social control takes place primarily through the enforcement of informal norms, as opposed to law.

Policing Disorder and "Broken Windows"

Governing crime through "broken windows" policing came to public attention when the first evidence of the decline in crime in New York of the early 1990s became available. Media reports linked this drop in the crime rate to a broad police campaign to increase misdemeanor arrests for disorderly behavior and public order offenses.[64] In the context of crime and policing, "broken windows" developed as a metaphor to understand how perceived disorder can set off a chain reaction that weakens informal controls in neighborhoods, leading to additional disorder and more serious crime.[65] "Broken windows" strategies were soon adopted by police departments in the United States, and several jurisdictions implemented these strategies through widespread arrests for disorderly conduct.[66] Most notably, the New York City Chief of Police, William Bratton, adopted the concept of "broken windows" both metaphorically and literally and asserted, "If you peed in the street, you were going to jail. We were going to fix the broken windows and prevent anyone from breaking them again."[67] Bratton nearly doubled misdemeanor arrests, despite only a small increase in the number of complaints.[68]

"Broken windows" was one of several policing reforms in the 1980s and 1990s that emphasized community-based approaches to policing, often through a focus on the maintenance of order and the reduction of symbolic and physical disorder.[69] These reforms included loitering ordinances,[70] curfews,[71] policing of panhandling and graffiti, and tactics such as reverse stings and "stop and frisks,"[72] all part of the "new policing" that emphasized local and community "quality of life" approaches to crime prevention and security.[73] The police reasserted authority through physical presence on the street and, in many cases, widespread arrests. The underlying rationale was to prevent disorder and thereby allow informal community social controls to reemerge and enhance neighborhood quality of life.[74]

There was some early evidence to support "broken windows" strategies. The decline in the rate of crime in New York during this period was dramatic. Crime dropped across a wide array of offences, and that drop has persisted to the present day.[75] Homicide alone remains down over 80 percent from its peak.[76] More broadly, data collected across a wide range of US cities suggest that disorder is significantly linked to crime, a finding which could give credence to the potential efficacy of "broken windows" policing strategies.[77]

Yet, there is little support for the thesis that "broken windows" policing is itself the cause of crime reduction. Contrary to the claims of police departments, municipal governments, and police executive forums that tended to support this revolution in policing strategy, the empirical evidence from city-based trials and experiments has turned out to be weaker than expected. Rigorous empirical research casts doubt on whether "broken windows" policing efforts have a salutary effect on crime rates.[78] Researchers have also cast doubt on the empirical validity of the earlier studies of disorder and crime that underlie the "broken windows" approach.[79] Even where crime rates fell after implementing "broken windows" strategies, statistical evidence on their causal impact is largely inconclusive, mainly because these cities relied on several new strategies simultaneously, and it was consequently impossible to disentangle retrospectively their independent impact.[80] Media reports and academic analyses were most optimistic about the New York Police Department's vigorous "broken windows" strategy. The best available evidence, however, suggests that even in New York the decline in crime is not directly attributable to "broken windows" strategies. Time-series analyses over a ten-year period suggest a reversion to the mean after crack-related crime spikes.[81] Panel-based econometric comparisons across New York City precincts demonstrate that reduction in rates of crime cannot be attributed to aggressive policing; rather, the decline in crime began at the same rate prior to these interventions.[82] Within New York, the effectiveness of strategies that target discrete geographic locations, rely on vigorous policing, and hold individual precinct commanders accountable for crime rates in their areas is uncertain.[83]

Those who do find positive impact from changed policing strategies attribute some reduction in crime to the inclusion of location-based approaches within the policing repertoire.[84] These include community problem-solving efforts by the police force that focus on strategies that ask how victims and offenders have become engaged in situations that lead to crime. A systematic review of studies finds that focused, problem-oriented strategies in hot spots are the most effective approaches within situational policing efforts.[85]

Building on the findings regarding collective efficacy, some of the most elegantly designed research in criminology suggests that any relationship between "broken windows" policing and crime reduction is largely spurious.[86] What is significant is the presence of community cohesion through collective efficacy, even in cases of observed disorder.[87]

This is not to argue that disorder is never meaningful. Instead, building on systematic social observation research in Chicago, Sampson argues that to be successful, a strategy of "broken windows" policing requires an understanding of how social meaning is attached—or not—to any particular "broken window" by community residents.[88] This social meaning may then have real consequences for collective efficacy. In contrast to how "broken windows" has been translated into police practices—in particular, into harsher law enforcement—what is required, therefore, is a more comprehensive understanding of individual beliefs that acknowledges how, even in the same neighborhood, people differ significantly in how much weight they give to disorder as a problem. These perceptions are systematically shaped by residents' education, class, age, and social position. Race is also important; whites tend to see disorder as more of a problem than other groups, even when they live in the same environment.[89] If policing neighborhood disorder can be linked to reductions in crime, it needs to account for how individual and collective cognitions and community norms such as collective efficacy shape how any single broken window will be interpreted.[90] "Just as memory is dependent on context," Robert Sampson argues in summarizing the conclusions of this research, "so is what we 'see.' "[91]

What do we conclude? The evidence about "broken windows" policing strategies is at best ambiguous with regard to its impact on crime reduction and prevention. By extrapolation, strategies that reduce disorder and emphasize resilience in the wake of attacks may contribute only indirectly to the effectiveness of deterrence by denial. "Broken windows" approaches that do not grapple with community contexts of norms and social life have been found to be generally ineffective, whereas enhancing perceptions of procedural justice and deepening the well of collective efficacy in neighborhoods are linked to more successful outcomes. This suggests that deterrence-by-denial strategies may work through processes of legitimation (of law and of state officials) and processes of delegitimation (through social sanctions). These strategies reconfigure social risks and rewards by acting on the social meaning embedded in relationships with state officials and with community members.

CONCLUSION

Robust evidence from criminology undermines some of the conventional wisdom about deterrence of terror by denial and suggests important new areas for research about the effectiveness of different strategies that seek to deny would-be attackers the opportunity they seek. First, we can argue with considerable confidence that displacement should no longer be considered as a highly likely consequence of strategies of denial. Second, drawing on the evidence from studies that find diffusion of benefits, we put forward the proposition that the success of deterrence by denial is likely to be enhanced by three factors: uncertainty about the scope of denial strategies, heightened sensitivity to the probability of failure, and higher estimates of loss.

The results are equally promising for a second set of strategies—the focus of this chapter—that enhance community engagement and collaboration with law enforcement and link deterrence by denial with delegitimation of acts of terror. At the neighborhood level, a sense of collective efficacy is essential, as is a shared belief in procedural justice. There is strong evidence that enhancing community engagement is associated with a reduction in crime through the denial of opportunity. Strategies that focus on inducing shame in a would-be attacker for the violation of community norms and making alternative norms and narratives salient among community members also find some support. These community-based strategies increase the probability of failure by increasing the likelihood of social sanctions against those who would commit acts that are not only illegal but also normatively impermissible, and by legitimizing law-abiding norms and narratives while delegitimizing crime and violence.

Taken together, these strategies provide support for deterrence by denial, whether through physical target-hardening or through linked social processes of denial and delegitimation. The first set of strategies works on adjusting the physical environment while the second and third adjust the normative context in which would-be offenders operate. As we have seen, these strategies can operate at the level of the individual, at the level of social bonds, at the level of shared norms and narratives available to community members, and at the ecological level of collective efficacy and shared expectations for behavior.

Yet, the mechanisms that these strategies follow are very different, with important implications for policy. The first set of strategies—target-hardening through increased surveillance—works very differently than do the second and third sets, which seek to deny opportunities to would-be attackers by enhancing a sense of community efficacy to promote collaboration with law enforcement and by decreasing the social rewards of impermissible action through community engagement. Target-hardening can increase uncertainty and the likelihood of failure, but it can create "surveillance societies" that fracture communities and heighten suspicion and stereotyping. And, over time, divided communities may create the social conditions that enable acts of terror. In contrast, the last two norm-based strategies strengthen deterrence by denial at two levels: they deny opportunities by inducing law-abiding citizens to provide intelligence to police that can thwart attacks, and they deny social rewards by creating a moral context that ensures offenders lose community esteem if they engage in socially prohibited activity. By reducing the social rewards to those who commit acts of terror and creating a moral context of cooperation with law enforcement, these strategies deny social approbation and community support to would-be attackers.

The first strategy stands in tension to the others; and using different strategies simultaneously can create very difficult trade-offs. The challenge is to model the appropriate balance amongst them, fully recognizing that some of these trade-offs are unavoidable. We do not have strong theoretical models or robust evidence to help structure the trade-offs among strategies of deterrence of terrorism by denial. We must look beyond political science to find useful analogues and evidence, because we cannot do the controlled or experimental studies that would generate robust inferences, nor are the number of cases large enough to generate valid results. In this chapter, we have drawn from a closely relevant field of study, criminology, and to extract evidence that is relevant to deterrence by denial through cognitive strategies that stress legitimation and delegitimation. The next step is to muster the evidence that engages with the complementarities and trade-offs among the different strategies that fall under the broad rubric of deterrence by denial.

NOTES

1. This chapter was originally published as "The Social Psychology of Denial: Deterring Terrorism," by Janice Gross Stein and Ron Levi, in *New York University Journal of International Law and Politics* 47 (2015) and is reproduced by permission.
2. The definition, designation, and meaning of terrorism are all contested. Terrorism has been defined as "[t]he unlawful use or threatened use of force or violence against individuals or property in an attempt to coerce or intimidate governments or societies to achieve political, religious, or ideological objectives." North Atlantic Treaty Organization (NATO) Standardization Agency (NSA), *NATO Glossary of Terms and Definitions*, at 2-T-5 (2008). The Global Terror Database at START, which is a Center of Excellence for US Homeland Security, uses a very similar definition in its most inclusive database: "the threatened or actual use of illegal force and violence by a non-state actor to attain a political, economic, religious, or social goal through fear, coercion, or intimidation." *Data Collection Methodology*, National Consortium for the Study of Terrorism and Responses to Terrorism. Definitions vary and are contentious, including as to whether the target must be civilian or can include state agents. Lafree and Ackerman, "The Empirical Study of Terrorism," 347, 348–49; Lafree and Dugan, "Introducing the Global Terrorism Database." In drawing together these two fields, it is significant to note that the Global Terror Database and START are directed by a criminologist who is past president of the American Society of Criminology.
3. Wilner, "Deterring the Undeterrable," 3, 6 (analyzing deterrence by denial as including delegitimation).
4. Perrow, *The Next Catastrophe*, 1–13.
5. Alex Wilner says that deterrence by denial "functions by reducing the perceived benefits an action is expected to provide. Hardening national or infrastructural defenses . . . Raises the costs of attack by diminishing the probability that an adversary is likely to acquire his objective." Wilner, "Undeterrable," 6. Although the theoretical argument is framed in cost-benefit language, it is probability that is doing the work.
6. Gross Stein, "Deterring Terrorism, Not Terrorists," 46.
7. Szyliowicz, "Aviation Security."
8. Sandler, "Collective Versus Unilateral Response," 75, 76.

9. *BBC News*, "Three Guilty of Airline Bomb Plot."
10. *Telegraph* "Underwear Bomber Plot Failed."
11. Lafree and Hendrickson, "Build a Criminal Justice Policy," 781.
12. Lum and Kennedy, *Evidence-based Counterterrorism Policy*; Clarke and Newman, *Outsmarting The Terrorists*; Lafree, "Expanding Criminology's Domain," 1 (advocating the introduction of a "wider emotional range" in dealing with terrorism and extending criminology principles to this area).
13. Gross Stein and Levi, "Testing Deterrence by Denial" (forthcoming).
14. Weisburd, et al., "Does Crime Just Move Around the Corner?", 550; Bowers, et al., *Systematic Review Protocol*.
15. Farrell, Chenery, and Pease, "Consolidating Police Crackdowns"; Grogger, "The Effects of Civil Gang Injunctions," 69.
16. Analyzing the potential of crime displacement necessarily requires attention not only to geography but also to time: without an experimental design, any apparent increase or decrease in crime may be a product of secular changes in crime rates that are unrelated to the intervention. See generally, Cantor and Land, "Unemployment and Crime Rates," 317, 323–324; Weisburd, et al., "Displacement of Crime."
17. Bowers, et al., *Systematic Review Protocol*, 3. Some may suggest that this is a product of specific deterrence, with offenders arrested as part of the intervention in the target site. Yet these findings persist even with interventions that do not rely on incapacitating offenders. Furthermore, in those cases where some crime does displace, that displacement is measurably less than the treatment effect, so that there remains an overall protective benefit from the intervention. Ratcliffe and Breen, "Crime Diffusion and Displacement," citing John E. Eck, "The Threat of Crime Displacement," Criminal Justice Abstracts 25 (1993); Guerette and Bowers, "Assessing the Extent of Crime Displacement," 1331, 1332.
18. Stein and Levi, "Testing Deterrence," (forthcoming).
19. Joel Caplan, et al., "Police-Monitored CCTV," 255; Farrington, et., al., "The Effects Of Closed-Circuit Television," 22, 33; Ratcliffe, et al., "The Crime Reduction Effects," 746; La Vigne, et al., "Evaluating The Use Of Public Surveillance Cameras."
20. Wilner, "Undeterrable," 26.
21. "Collective efficacy" refers to communities that can achieve their goals collectively.
22. Sampson, et al., "Neighborhoods and Violent Crime," 918.
23. Sampson, "The Place of Context," 20.

24. Tucker, *Strategies for Countering Terrorism*. On responses to terrorism, see generally, Spilerman and Stecklov, "Societal Responses to Terrorist Attacks," 167.

25. Tyler, "Toughness Vs. Fairness," 353, 354–355.

26. Roach, "For the Sake of Civility."

27. Tyler, *Why People Obey The Law*; Tyler and Fagan, "Legitimacy and Cooperation," 231, 264–265.

28. Tyler, "What Do They Expect?", 22; Tyler, "Legitimacy and Criminal Justice," 307, 319–322.

29. Tyler and Huo, *Trust in the Law*, 49–57; Tyler, *Why People Obey*, 3; Tom Tyler, "Restorative Justice And Procedural Justice," 307, 309–313.

30. Tyler and Fagan, "Legitimacy and Cooperation," 267.

31. Ibid., 248.

32. Ibid., 249, 252.

33. Huq, et al., "Mechanisms for Eliciting Cooperation," 728, 750; and Tyler, et al., "Legitimacy and Deterrence Effects," 365, 380, Table. 1.

34. In Britain, procedural justice has a direct effect on cooperation, but is not mediated through "legitimacy," leading the authors to hypothesize that cooperation in the United Kingdom is more contingent—based on personal treatment—than based on a broader legitimacy of the policing institution or the state. Huq, et al., "Mechanisms for Eliciting Cooperation," 728, 750.

35. Tyler, et al., "Legitimacy and Deterrence Effects," 365, 380, Table. 1.

36. Huq, et al., "Why Does the Public Cooperate," 419.

37. Tyler, "Legitimacy and Deterrence Effects," 381–387; Huq, et al., "Mechanisms for Eliciting Cooperation," 750.

38. Innes, "Policing Uncertainty," 222.

39. Huq, et al., "Mechanisms for Eliciting Cooperation," 750.

40. Wortley, "Guilt, Shame, And Situational Crime Prevention," 115, 123–127.

41. Cronin, *How Terrorism Ends*, 71.

42. Khalid, "Some Experts See Fatwa."

43. See Moghadam, "Motives for Martyrdom," 46, 60–61 (Noting the Example of Osama Bin Laden Praising Young Muslims Who Commit Violence as Martyrs Who Will Prevail Against the "Crusaders").

44. Sampson, et al., "Beyond Social Capital," 633.

45. See Kennedy, *Don't Shoot*; David Kennedy, *Deterrence And Crime Prevention*.

46. See Kennedy, "Practice Brief," (discussing the practical aspects of addressing "norms and narratives" in crime prevention). See also Kennedy, "Drugs, Race And Common Ground," 12–17, (discussing the mobilization of "influentials" and others who have close relationships with drug dealers to encourage them to stop).
47. See Kennedy, "Practice Brief," (on encouraging this type of dialogue between "influentials" and drug dealers).
48. Hirschi, *Causes Of Delinquency.*
49. Kennedy, "Practice Brief,"11.
50. Kennedy, Presentation at The 5th Annual Restorative Justice Initiative Conference.
51. Kennedy, *Don't Shoot,* 69.
52. Kennedy, *Deterrence*; Kennedy, "Drugs, Race And Common Ground," 17.
53. Cook, "The Impact of Drug Market," 161, 162.
54. Braga, "Getting Deterrence Right? ", 201, 205. See also Braga and Weisburd, "The Effects Of "Pulling Levers," 1.
55. Bowers, 63–69, table 4; Braga and Weisburd, "The Effects Of "Pulling Levers," 37.
56. This has been defined by The National Research Council. Skogan and Frydl, *Fairness And Effectiveness In Policing* ("The heart of problem-oriented policing is that this concept calls on police to analyze problems, which can include learning more about victims as well as offenders, and to consider carefully why they came together where they did. The interconnectedness of person, place, and seemingly unrelated events needs to be examined and documented. Then police are to craft responses that may go beyond traditional police practices .... Finally, problem-oriented policing calls for police to assess how well they are doing. Did it work? What worked, exactly? Did the project fail because they had the wrong idea, or did they have a good idea but fail to implement it properly?").
57. This research returns to the sociology of the Chicago School of the 1920s that emphasized the importance of neighborhood context but lapsed into ecological fallacies about the effects of these contexts for individual outcomes. This work was instrumental in introducing the use of new methodological tools such as hierarchical linear modeling (HLM) developed by Raudenbush in the early 1990s (which can account for variance across levels). Raudenbush and Byrk, *Hierarchical Linear Models,* 459–496. Collective efficacy is a latent construct understood as the linking together of social cohesion among residents combined with shared expectations over a community-based willingness to intervene on behalf

of the common good, whether to monitor children, respond to disorder, intervene to stop violence, or secure city services. Questions about collective efficacy have become standard in the field, focusing on the willingness of neighbors to intervene if children were skipping school and hanging out on a street corner, if children were spray-painting graffiti on a local building, if children were showing disrespect to an adult, if a fight broke out in front of their house, and if the fire station closest to their home was threatened with budget cuts. Sampson, et al., "Neighborhoods and Violent Crime."

58. Sampson, "The Place of Context," 20. Survey research has been augmented through systematic social observations across Chicago over a five-month period. The systematic social observations involved videotaping and observing across all of Chicago's census tracts over a five-month period, then sampled and coded for nearly 24,000 "face blocks" (one side of a city block) across the city. Morenoff, et al., "Neighborhood Inequality, Collective Efficacy," 517; Sampson and Raudenbush, "Systematic Social Observation Of Public Spaces," 603, 616.

59. Sampson, et al., "Neighborhoods and Violent Crime," 18–24.

60. Jonathan Jackson, "A Psychological Perspective on Vulnerability," 365; Jackson, et al., *Just Authority?*. The ecological literature also draws attention to neighborhood contexts of legal cynicism, where violence is higher where legal rules are perceived as irrelevant. This cognitive landscape can be reinforced by disadvantage, with a feedback loop that feeds further cynicism and behavior outside the law. Research demonstrates that such cynicism helps explain the persistence of violence. There are some preliminary indications in the United Kingdom that neighborhood level cynicism may itself lead to lower levels of cooperation with police. In the United States, arrests for crimes are less likely in neighborhoods with higher levels of legal cynicism. For deterrence by denial, then, we need to consider the elements that are protective against legal cynicism, and predictive of collective efficacy. Kirk and Papachristos, "Cultural Mechanisms," 1190; Kirk and Matsuda, "Legal Cynicism, Collective Efficacy," 443, 460; Hough, et al., "Procedural Justice, Trust," 203, 207; Sampson, "The Place of Context," 18n16; Sampson, "Moving And The Neighborhood Glass Ceiling," 1464, 1465; Sampson and Bartusch, "Legal Cynicism And (Subcultural?) Tolerance," 777.

61. Weisburd, et al., "Understanding And Controlling Hot Spots."

62. Meares and Kahan, "Law and (Norms Of) Order," 805.

63. Sampson, *Great American City*, 414–426.

64. Livingston, "Police Discretion," 551, 578.
65. Kelling and Wilson, "Broken Windows."
66. Eck and Maguire, "Have Changes in Policing Reduced Violent Crime?"; Harcourt, *Illusion of Order.*
67. Bratton and Knobler, *The Turnaround,* 229
68. Harcourt, "Reflecting On The Subject," 291, 340.
69. Rosenbaum, *The Challenge of Community Policing*; George Kelling and Mark Moore, "The Evolving Strategy of Policing"; Livingston, "Police Discretion."
70. Levi, "Making Counter-Law," 131, 132.
71. Meares and Kahan, "Law and (Norms Of) Order," 832.
72. Kahan, "Between Economics and Sociology," 2477, 2488; Eligon, "Taking On Police Tactic."
73. Livingston, "Police Discretion."
74. Skogan, "Broken Windows," 195, 198–200.
75. Eck and Maguire, "Have Changes in Policing Reduced Violent Crime?", 224–225; Greenberg, "Studying New York City's Crime Decline." See also Zimring, *The City That Became Safe* (analyzing how and why the crime rate in New York dropped).
76. Zimring, *The City That Became Safe,* 5.
77. Skogan, *Disorder and Decline* (finding close links between the prevalence of social and physical disorder and crime rates).
78. Greenberg, "Studying New York City's Crime Decline," 155; Harcourt, *Illusion of Order*; Harcourt and Ludwig, "Broken Windows," 271.
79. Harcourt, "Reflecting on The Subject," 302–305 (critiquing the hypothesis that minor physical and social disorder causes serious crime if left un-policed). In his proposal reviewing the research in this field, Anthony Braga summarizes that "evaluations of the crime control effectiveness of broken windows policing strategies also yield conflicting results," and that "[i]n New York City, for example, it is unclear whether broken windows policing can claim any credit for the 1990s crime drop." Anthony Braga, "Title Registration." He notes that in New York City, for example, it is unclear whether broken windows policing can claim any credit for the 1990s crime drop. See Karmen, *New York Murder Mystery* (discussing potential explanations for decreases in crime during 1990s); Eck and Maguire, "Have Changes in Policing Reduced Violent Crime?", 226 (surveying scholarly claims that police tactics have varying degrees of responsibility for decreases in crime). Braga notes that some evaluations report significant reductions in violent crime following broken

windows policing. See Corman and Mocan, "Carrots, Sticks, And Broken Windows" (investigating the validity of the broken windows hypothesis); Kelling and Sousa, "Do Police Matter?" (concluding that policing is "significantly and consistently linked to declines in violent crime"). Others report modest reductions in violent crime. Messner, et al., "Policing, Drugs," 385; Rosenfeld, et al., "The Impact Of Order-Maintenance Policing," 355. Harcourt and Ludwig, in "Broken Windows," report no evidence of reductions in violent crime. Braga concludes that "[t]hese conflicting results have generated questions on the crime prevention value of dealing with physical and social disorder." Braga, "Title Registration." A meta-analysis of studies finds that focused strategies that are problem-oriented in hot spots are most effective. Weisburd and Eck, "What Can Police Do," 42, 60.

80. Greenberg, "Studying New York City's Crime Decline," 155–156.
81. Harcourt and Ludwig, "Broken Windows," 276.
82. Greenberg, "Studying New York City's Crime Decline," 155.
83. Zimring, *The City That Became Safe*, 142–147; Eck and Maguire, "Have Changes in Policing Reduced Violent Crime?", 230–235; Greenberg, "Studying New York City's Crime Decline."
84. Weisburd and Eck, "What Can Police Do"; David Weisburd, et al., "Policing, Terrorism," 203.
85. Braga, Papchristos, and Hureau, "The Effects of Hot Spots," 633. The national decline in crime rates also suggests that the drop in crime in New York was also a function of secular trends that cannot be attributed to specific changes in the city's policing strategy. Researchers have pointed to changes in drug markets and gang consolidation, demographics, incarceration policies, increased surveillance, or changes in youth culture. Rosenfeld et al., "The Impact Of Order-Maintenance Policing." The question of whether increased numbers of police officers may be at the root of crime changes is itself very difficult to parse methodologically, largely because of causation problems (since changes in police strength may result from changes in crime rates). The evidence here is ambiguous, though the most sophisticated research suggests that there may be some decrease in crime rates achieved by increasing police strength. Weisburd and Eck, "What Can Police Do," 49.
86. Sampson and Raudenbush, "Systematic Social Observation of Public Spaces."
87. Where disorder has been found to be directly linked to crime, it is limited to robbery, rather than to crime in general. Sampson and Steve Rauden-

bush, "Disorder in Urban Neighborhoods." Similarly, in an experimental study of five US cities where low-income residents were provided with "moving to opportunity" housing vouchers, there was no net reduction in crime for residents who were assigned to move to less disorderly neighborhoods, suggesting at best an ambiguous relationship between the presence of disorder and effects on individual crime as well as the importance of attending to other neighborhood characteristics. Harcourt and Ludwig, "Broken Windows."

88. Sampson, "The Place of Context," 16.
89. Ibid.; Hipp, "Resident Perceptions," 475, 479.
90. Sampson, "The Place of Context," 17.
91. Ibid., 16.

CHAPTER 4

DISSUASION BY DENIAL
IN COUNTERTERRORISM

THEORETICAL AND EMPIRICAL DEFICIENCIES

John Sawyer

The intent of this chapter is to explore current theoretical and definitional deficiencies in the application of deterrence by denial to counterterrorism and to suggest a way forward with the development of a new concept—dissuasion by denial. While traditional international security concepts and strategies, including those proposed and developed in this volume (e.g., Patrick Morgan, Alex Wilner, James Wirtz, Jonathan Trexel, Dima Adamsky) are generally applicable to counterterrorism, there are key differences that need to be addressed. States face a wide variety of terrorist adversaries, ones who range across the ideological spectrum; possess various degrees of size, resources, hierarchy, territoriality and/or popular support; and engage in behaviors beyond terrorism (e.g., providing public services). Furthermore, these attributes change over time, often in highly dynamic ways. As such, insights from traditional IR theory may be applicable to terrorist adversaries when these adversaries approximate another state, while insights from criminology (as highlighted by Janice Gross Stein and Ron Levi in chapter 3 of this volume) may be more applicable to terrorist adversaries that are small, weak, and lacking in territorial assets. However, an amalgamation of approaches is necessary for terrorist groups that spend any time in the intermediate range of the spectrum. To further complicate things, states are often playing multilevel counterterrorism games: facing multiple terrorist threats while also dealing with politics in the international and domestic spheres. Actions directed at one audience or adversary have implications for the others.

To date, there has been insufficient attention to mapping the conceptual issues underlying deterrence by denial in light of the complexity of the counterterrorism challenge, which has significant implications for the ability to empirically evaluate such policies in any meaningful way. Beyond impeding such an academic exercise, the lack of conceptual clarity has the potential to lead to muddled and suboptimal policies to counter the threat of terrorism. Therefore, much of this chapter is dedicated to a theoretical discussion of the logics of counterterrorism to distill a more coherent concept of preventative influence (e.g., dissuasion by denial). Illustrative examples from Israel and Northern Ireland are then used in the conclusion to highlight next steps for terrorism and deterrence scholarship.

LOGICS OF COUNTERTERRORISM

Although the concept of counterterrorism has grown over the years to include a wide range of tactics and approaches, there are two funda-mental goals to be considered in developing a counterterrorism strategy: 1) mitigating the effects of terrorist attacks when they occur, and 2) preventing the adversary from carrying them out in the first place.[1] Given the variety of means to accomplish these goals, the specific attributes of the terrorist threat, and the need to weigh these counterterrorism goals against other societal imperatives, there is no one-size-fits-all counterterrorism strategy. It is therefore essential to have a flexible but coherent framework to evaluate the rationale and likely effects of a counterterrorism policy within a given context. The framework provided in this chapter builds on a model of terrorism/counterterrorism as a dynamic interaction between one or more attackers (or challengers) and one or more defenders, each with a variety of actions available to them to pursue divergent interests.

If one assumes that both defender and attacker have some basic rationality, their interactions are defined by some kind of cost-benefit analysis given the perception of the other's likely behavior.[2] While both the defender and attacker have their own decision calculus for defining optimal strategies, this chapter will focus primarily on the adversary's decision-making—and opportunities for its manipulation. An adversary's decision calculus is characterized by an estimation (with some degree of uncertainty) of the costs and benefits of both action and inaction.[3] The adversary's costs include the consumption of resources by the action itself or from natural attrition, the risk of larger organizational losses, and negative perceptions within its constituent base. The adversary's benefits include acquiring leverage over the defender, enhancing the organization's capabilities, and positive perceptions within the pool of potential recruits or supporters (including current members). However, it is critical to recognize that given resource constraints, an adversary cannot choose a behavior whose costs exceed available resources no matter how beneficial the behavior would be.[4]

As such, there are three basic, potentially overlapping approaches to achieving these counterterrorism goals, each of which targets a different part of the adversary-defender strategic interaction: *defense* focuses on the defender's vulnerabilities and costs; *offense* focuses on the adversary's capabilities; and *influence* focuses on the adversary's perceptions of likely costs and benefits.

This trifurcation is not absolute, and the three approaches are generally closely linked. Defense and offense are rather distinct, and operations usually fall into one category or the other. However, given the fluidity of concepts like combatant status in the realm of counterterrorism, some operations potentially blur the lines between them (e.g., a preemptive strike on a terrorist cell known to be actively plotting an attack). In contrast, influence only rarely occurs in isolation from at least one of the other two approaches. On the one hand, the defender's decision to invest in either offensive operations or defensive capabilities generally include an assessment of the risk, especially the likelihood of an attack occurring; this likelihood can be directly affected by investments in influence efforts. On the other hand, adversaries often factor in the offensive and defensive capabilities they know about or suspect when planning and executing operations; indeed, this adaptation to the defender's moves is one of the core elements of the influence approach. Unfortunately, the large number of options makes this adaptation hard to predict (e.g., temporary or permanent inaction, target shifting, tactic shifting, resource shifting, goal shifting). Moreover, many influence operations involve a threat or a promise of (in)action and rely on the credibility derived from past (in)action. Given the broad definitions explored in the following section, many parts of the current conceptualization of deterrence by denial lie squarely at the nexus of these three approaches.

In sum, each approach centers on affecting a different part of the strategic adversary-defender interaction. Defense concentrates almost exclusively on the defender's cost calculus with little consideration of the adversary's responses. Offense seeks to reduce an adversary's resources to the extent that the costs of action exceed its capacity irrespective of the benefits the adversary would accrue from such action. Influence seeks to alter the adversary's perceptions of the likely costs and benefits of (in)action. Influence operations may be narrowly aimed at a particular component of the adversary's cost-benefit calculus, or they can be quite broad and introduce new opportunity costs or side-benefits that fundamentally alter that calculus. A discussion of each approach is provided next.

Defense

In the counterterrorism context, defense is the ability to decrease the casualties, disruption, and political impact of terrorism. Defense is actively target-focused and is characterized by a relatively high certainty about its effect. In other words, defense is a unilateral effort to reduce the costs to the defender in the event an adversary attempts an attack against a given target and is not directly aimed at affecting the adversary's decision calculus. For example, blast mitigation barriers are certified to absorb a given amount of force. Similarly, investments in community resilience reduce the disruptive effects of a disaster. These investments are primarily intended to reduce damage and save lives, not to prevent an event from occurring in the first place.

There are three primary logics of defense: protection, mitigation, and resilience.[5] Protection consists of efforts to put the target beyond the reach of a potential adversary: creating layered active or passive defenses around a target to stop an adversary from approaching and attacking it. Mitigation consists of efforts to reduce the damage from an attack: designing structures and systems to protect personnel or key resources or reducing the normative or symbolic value of the target. Resilience consists of efforts to reduce the significance to the defender's ongoing operations of the target's loss if it is attacked: building redundant systems. It is heuristically useful to consider these defensive logics together as concentric rings around the target (see figure 1). If an adversary attacks, protection will prevent him from reaching the target. If an adversary is able to penetrate the protection, mitigation will minimize the resulting damage. And if the target is critically damaged, the defender will be able to replace the lost resources quickly.

Figure 1. Logics of Defense.

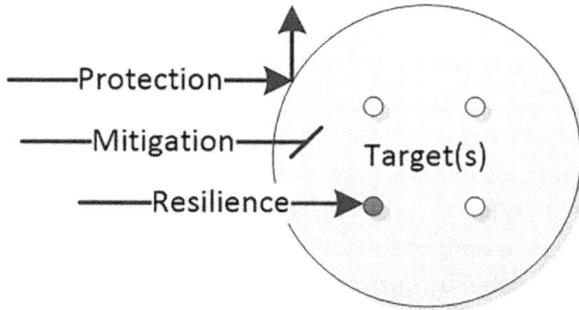

While these defensive logics are often complementary, they can nonethe-less be employed singly and may at times be contradictory. For illustra-tion of the latter, efforts to build resilience by creating redundant assets could increase the number of targets that must be protected, and/or make it harder to preposition responder resources that could mitigate the effects of an attack. Likewise, some efforts to protect a given target may reduce the ability to mitigate or bounce back from a successful attack. For example, large public investments in airport screening may provide significant protection, but they also increase the political impact when terrorists manage to get through—even if the attack itself is unsuccessful.

Thus, defense assumes that at least one potential adversary exists and will at some point attempt to carry out an attack but seeks to minimize the negative impacts of such an attack. However, given the resource constraints on the defender, these investments must clearly take into consideration assessments of the implementation costs, including its impact on the defender's other key functions and normative values, and the likelihood of an adversary attacking.

Offense
Offense includes any effort that reduces a terrorist's capacity to plot and carry out an attack on an adversary. Offense takes the battle to

the enemy, attempting to eliminate his capacity to fight in part or in whole. In other words, it is primarily focused on increasing the resource constraints to restrict or eliminate an adversary's behavioral options. As a bilateral interaction with asymmetric information, it is more dynamic and its effects are often more uncertain than defensive efforts.

Offense draws heavily on the coercive concept of preemption: removing an adversary's ability to make a decision to act or not by making it impossible for the adversary to act at all.[6] While kinetic operations that eliminate specific terrorists may be the most obvious example, they are the tip of the iceberg. Indeed, most traditional counterterrorism activities fit squarely into offense (e.g., elimination of safe havens, combating terrorist financing, arrest of suspected members). Given this chapter's (and the volume's) focus on deterrence by denial, which lies primarily at the intersection of defense and influence, it is unnecessary to fully explore these logics individually.[7]

One exception is that efforts to dry up the pool of potential recruits through delegitimization must be more clearly defined. Contrary to the treatment by some scholars, efforts to delegitimize an ideology, key individuals or an organization fit more appropriately within this offensive logic rather than a distinct sub-type of deterrence.[8] However, efforts to delegitimize a specific behavior, like targeting civilians, are well within the domain of influence as discussed later. For example, efforts to undermine the appeal of al Qaeda by citing its perversions of Islamic doctrine aim to restrict the recruitment pool generally, while efforts to delegitimize al Qaeda by citing the large number of Muslims killed in their attacks aim to force a behavioral change away from indiscriminate violence. Admittedly, these two forms of delegitimization may be difficult to disentangle because perceptions about actors and their actions, intentions and environments are generally not independent.[9] The key is that the former is clearly aimed at raising the costs of doing business for a particular adversary whereas the latter is aimed at affecting the adversary's cost calculation about a particular behavior.

In summation, offense is squarely focused on reducing or eliminating the resources adversaries have at their disposal to limit the behavioral options available to them. As with defense, offensive options must be weighed within a larger decision framework, which should include legal and normative costs. Thus, whereas genocide of an entire constituent population may eliminate the threat from a particular terrorist group, it would be both morally reprehensible and fraught with legal and political consequences. It is therefore unsurprising that the empirical record is replete with instances of more restrained counterterrorism policies designed to limit adversaries' options rather than destroying them outright.

Influence

Influence is the ability to have an effect on the nature or behavior of an adversary beyond simple coercion. Influence is often indirectly adversary-focused, leveraging strategic interactions between multiple entities within the larger sociopolitical environment to maintain or alter the adversary's behavior. Influence is thus an attempt to get inside the adversary's decision calculus and manipulate its perceptions of the costs and benefits. Although conceptually distinct, it is often difficult to differentiate between the physical acts intended to change these costs and benefits and the communicative acts intended to change perceptions of these costs and benefits—often the same act is supposed to perform both functions.

Given the multilateral interactions involved, influence is highly dynamic and characterized by relatively high uncertainty. The defender has enormous agency in both offense and defense (i.e., it has fairly well-defined objectives, can choose when and where to take action and can estimate the effects of such actions with relative ease and certainty). In contrast, a defender's agency for influence is filtered and mediated by many factors. As such, the focus is on the often difficult task of communication to potential adversaries of the consequences of their (in)actions so they will choose behaviors more closely in line with the defender's preferences.[10] Thus, the key element to influence is that the adversary retains a degree of volition rather than being forced into compliance or being eliminated altogether.

Paul Davis and Brian Michael Jenkins identify a wide range of coercive and non-coercive ways to influence current and potential adversaries at multiple different levels: Co-optation, positive inducements, persuasion, dissuasion, and six forms of deterrence (by threat, risk escalation, denial, punishment, defeat, and crushing).[11] These boil down to actions that affect an adversary's estimation of the costs and/or benefits of an action (including opportunity costs and benefits of inaction), or the nature of the cost-benefit equation itself. As an example of the former, the installation of metal detectors in airports in 1973 increased the likelihood of terrorists being arrested before they could hijack a plane, simultaneously raising the costs and lowering the benefits of this type of attack. As Walter Enders, Todd Sandler, and Jon Cauley demonstrate, this led to a reduction in terrorist hijackings.[12] Whereas this is primarily a defensive action designed to protect aviation, the public visibility of the installation and its effects made it easy for adversaries to update their assumptions about the costs of (some) aviation attacks. As an example of the latter, the creation of the Palestinian Authority (PA) fundamentally altered the cost-benefit equation for Fatah by legitimating its political activities. This political structure provided a substitute to violence for advancing its goals and a new set of potential costs to weigh in decisions about engaging in terrorism.

Likewise, Davis and Jenkins highlight that counterterrorism actions are often simultaneously directed against an array of actors with varying goals, values, and structures, which can lead a given action against one adversary to have a number of positive or negative influence externalities on the others.[13] Building on the previous example, the creation of the PA was primarily directed at eliminating the terrorist threat from the PLO, but it had massive ripple effects on other Palestinian groups, most notably Hamas. Similarly, the Global War on Terrorism had a number of successes in reducing the capabilities of al Qaeda and its affiliates but also led to the growth of other affiliates and domestic radicalization of a number of individuals.

While influence seeks to alter the adversary's expected utility calculation, there are two basic logics by which this can be accomplished. Arguably, insufficient attention to the differences between these two logics has had a profound effect on the conventional naming and understanding of deterrence by denial. The first logic is "bundling," in which an adversary's (in)action is linked to a defender's (in)action that will eliminate any expected utility from the initial (in)action. Under this logic, the adversary is the first mover in a game whose rules the defender has defined but whose enforcement potentially suffers from uncertainty and credibility issues. The second logic is what can generally be called "dissuasion," in which the defender takes actions that would make an adversary's undesired actions appear unprofitable.[14] The defender makes the first move and the adversary must adjust its cost-benefit analysis to include these changes. The logic of bundling introduces exogenous considerations (costs and/or benefits outside the status quo) into the adversary's decision calculus, whereas dissuasion seeks to alter perceptions of the direct costs and benefits of the targeted behavior. In other words, the former focuses on a strategic interaction between the defender and adversary; the latter focuses on manipulating general perceptions of the strategic environment.

INFLUENCE AS BUNDLING

Bundling is the linkage of two distinct actions in an attempt to force an adversary to consider the joint utility of their action and the defender's reaction.[15] In essence, this introduces additional, conditional cost and benefit components into the adversary's calculus of the utility of action versus inaction. Table 1 highlights the key features of three forms of bundling: compellence, deterrence, and inducement, all of which can have positive and negative variations. Positive versions seek to encourage the adversary to engage in a tangible behavior; negative versions seek the prevention or desistance of a behavior.

Either way, the logic of bundling includes two pieces of information that must be communicated to a potential adversary. The first element is the threat of diminished utility (from punishment "Y" or lack of positive inducement "Z") if the adversary fails to comply with the demand; the second element is the promise of non-action or payment if the adversary does comply. As Davis points out, the second element can pose a major problem for the applicability of deterrence to the counterterrorism puzzle.[16] Whereas this logic can work well for deterring an individual or group from the initial decision to engage in terrorism—keep being a law-abiding citizen and the government will not attempt to arrest or kill you—with the exception of amnesties and pardons, it works far less well for subsequent decisions—stop engaging in terrorism and we will still punish you for past transgressions.

Table 1. Logics of Bundling.

	Positive	Negative
Deterrence	Keep doing X and I will NOT do Y; Stop doing X and I will do Y	Keep NOT doing X and I will NOT do Y; Do X and I will do Y
Compellence	Start doing X and I will NOT do Y; Keep NOT doing X and I will do Y	Stop doing X and I will NOT do Y; Keep doing X and I will do Y
Inducement	Do X and I will do Z; Keep NOT doing X and I will NOT do Z	Do NOT do X and I will do Z; Do X and I will NOT do Z.

INFLUENCE BY DISSUASION

Dissuasion involves communicating information to an adversary that could alter his perception of the status quo costs and benefits associated with the execution of a particular behavior and encourage him to choose a different course of action. Thus, dissuasion is distinct from bundling because of its focus on the environmental status quo rather than conditional promises or threats to introduce additional benefits or costs. The most obvious form of dissuasion focuses on changing the cost calculation for a given behavior by transmitting or suppressing information about the quality and nature of the protection around a given target, the susceptibility of certain types of behavior to a defender's offensive operations, and the likelihood of detection and interdiction. Equally important are efforts to adjust the expected benefits to be derived from the behavior. If the defender and adversary have inverse preferences—adversaries seek to maximize the damage to the defender and the defender seeks to minimize its damage—the publication of mitigation and resilience efforts would reduce the expected benefit of an attack.[17]

While the examples discussed earlier focus on the connection with offensive or defensive operations, dissuasion can occur independently. As previously mentioned, deterrence by delegitimization at the tactical level could also prove to be a potent tool of dissuasion by increasing the adversary's perception of the costs of that behavior on levels of constituent support. Similarly, public diplomacy campaigns designed to soften the negative image of the defender held by the adversary and its constituents, especially vis-à-vis other potential targets, could significantly reduce the relative benefits of an attack. Alternatively, as has been done with Chemical, Biological, Radiological and Nuclear (CBRN) terrorism, the defender can publicize sufficient details of highly complex processes to demonstrate the prohibitive costs necessary to pursue an undesired behavior or capability.

RETHINKING DETERRENCE BY DENIAL FOR COUNTERTERRORISM

The current volume is dedicated to exploring the concept of deterrence by denial, writ large and across the domains of conflict, but this chapter urges a reconsideration of both the term and the concept, at least within the context of counterterrorism. Current definitions of deterrence by denial appear to include a mishmash of the bundling, dissuasion, and offense logics, but empirical studies tend to focus on defenders' defensive behaviors. While the inclusion of "deterrence" in the name suggests an emphasis on the logic of bundling, the core component of strategic target hardening generally falls within the logic of dissuasion.[18] Therefore, this chapter introduces the more narrowly focused concept of *dissuasion by denial*, defined as efforts to change an adversary's tactical behaviors by manipulating perceptions of the ability to access and attack a given target using a given tactic, the immediate and long-term costs of attempting such an attack given the defender's prior actions, and benefits to be accrued vis-à-vis the defender or the adversary's constituency. The concept is expanded upon next.

Deterrence vs. Compellence
There have been extensive discussions about the ability and desirability of deterring terrorism; indeed, it is a longstanding cornerstone of US counterterrorism policy.[19] However, the discussion to date on counterterrorist deterrence has not adequately addressed the significance of the distinction between deterrence and compellence. With the exception of deterring recruitment or certain extreme behaviors like CBRN terrorism, much of what has been suggested targets behaviors that have already been adopted by groups or individuals. As such, the strategy is really about compellence. While this may appear to be an issue of semantics, Schelling notes that compellence is a far more difficult strategy to pursue effectively because it requires an adversary give up a manifestly preferred behavior into which it has already invested.[20]

This mislabeling is quite understandable given the influence of criminological definitions of deterrence on the discussion; however, this concept has been muddled in its application. Criminological deterrence focuses on efforts that reduce the offending—especially recidivistic offending—rates of individuals, a theme explored in Stein and Levi's chapter to this volume.[21] The criminal justice system is oriented around two distinct forms of deterrence: general and specific.[22] General deterrence—the public punishment of deviants to increase the credibility of threatened sanctions to others by increasing perceptions of the celerity, severity and certainty of punishment—has direct applicability for counterterrorism but falls within the logic of dissuasion. The punishment terrorists receive doubtless sends a very strong signal about status quo costs to non-terrorists considering a change in behavior. Similarly, the perception of past adversaries' ability to penetrate defenses, overwhelm mitigation or resilience efforts, and elude punishment will provide a trusted source of information for potential adversaries' estimation of the costs of current action.

In contrast, the criminological concept of specific deterrence—punishments intended to deter an offender from committing the next deviant act—is of much more limited value in the counterterrorism toolkit. Most criminals primarily engage in legitimate behaviors punctuated by spurts of illegal activities; deterrent influences must therefore keep criminals anchored in their prevalent legal behaviors in order to prevent recidivism.[23] In contrast, under most Western legal systems, once an organization is labeled "terrorist" all of its activities are illegal. Similarly, individual terrorists' main protection from punishment for past misdeeds is continuing membership in the organization, which entails engaging in a host of illegal activities.[24] Therefore, rather than deterring a previously offending organization or individual from re-engaging, "deterrent" counterterrorist policies are really attempting to compel them to change their behavior.

Inconsistent Logics of Denial at Different Levels of Analysis

Although the literature has done an excellent job in addressing the primary conceptual challenge of deterrence by denial—differentiating between the self-centered logic of defense and adversary-oriented logics of deterrence by denial—current definitions appear to include an untenable and confusing mishmash of the bundling, dissuasion and offense logics discussed previously.[25] Traditional conceptualizations of deterrence by denial have revolved around the communicative elements of target-hardening efforts,[26] but work over the last decade has dramatically increased the conceptual scope and pushed it to the breaking point. This expansion of the concept has been captured in Smith and Talbot's influential framework, which highlights the potential applicability of deterrence by denial beyond denial of opportunities at the tactical level to include denial of capability at the operational level and denial of objectives at the strategic level.[27]

Tactical denial—communication about target-hardening efforts— conveys to the adversary that it must adopt new tactics, which are likely to increase the cost or uncertainty of action in the short term, accept an increased probability of failure using the old tactics, or forgo the target entirely if it is either no longer politically profitable or beyond its resource constraints.[28] Given the intent to preemptively alter the adversary's perception of the costs and benefits, this is a clear case of dissuasion.

Capability denial, when defined as restricting the adversary's access to resources required for the targeted behavior, is largely indistinguishable from an offensive strategy aimed at eliminating the adversary's capability. However, when it is defined as strategic communication intended to alter perceptions about the availability and acquisition costs for necessary resources (e.g., public dissemination of knowledge about the range and depth of the technical skills required to build a nuclear weapon), it too is a case of dissuasion. However, because the requisite resources for most terrorist attacks are fairly basic and highly fungible, capability denial is probably only a useful counterterrorism tool when directed against the highly sophisticated end of the attack spectrum. Even then, capability denial is arguably simply a subset of tactical denial—it is just directed at an earlier stage in a longer acquisition-attack process, wherein the protected resources are the targets for acquisition behaviors.

Finally, it has been argued that strategic denial—removing hope of ever achieving the adversary's strategic goals through the targeted behavior— may be the most important type of deterrence by denial.[29] Unfortunately, strategic denial is sufficiently amorphous to include several different logics. Most versions of this concept appear to rely on the logic of bundling: "if you do not stop engaging in terrorism, I will never accede to any of your goals; if you stop engaging in terrorism, I might accede to some of your goals." Some versions could rely on the logic of dissuasion by removing the conditionality of that threat to state that pursuit of the goal is itself futile, reducing the perceived benefit of all of the adversary's actions, not just the behavior being targeted. Both versions are limited by the defender's credibility in denying these goals in both the short and long term.

Given the logical ambiguities that attend this expansion of the concept of deterrence by denial to the operational and strategic levels, this chapter adopts a much more restrictive focus on the tactical level and is focused on the logic of dissuasion rather than bundling.

Dissuasion by Denial

Dissuasion by denial consists of efforts to alter an adversary's perceptions of the ability to access and attack a given target using a given tactic, the immediate and long-term costs of such an attack (including from status quo responses of the defender), and benefits to be accrued vis-à-vis the defender or the adversary's constituency. There are four types of denial communications: 1) defensive denial (target hardening); 2) behavioral denial (increasing environment uncertainty); 3) mitigation ("blunting and limiting terrorism's social, political, and economic effect"); and 4) behavioral delegitimization (increasing the constituency costs/reducing the benefits of certain tactics or targets).[30]

There is usually a relatively strong correlation between the objective strategic environment and the adversary's perception thereof; failure to align quickly with objective reality usually leads to swift termination. Observable changes in the defender's capabilities or behaviors are likely to influence the adversary's calculations irrespective of the defender's intent to alter the adversary's perceptions. However, many factors that affect the cost-benefit analysis may not be directly observable, giving the defender an opportunity to manipulate the adversary's perceptions of them.

As such, the first three types of communication are closely connected with the defender's concrete defensive or offensive actions. Defensive denial is the discursive extension of the logic of defensive protection; mitigation is the discursive extension of defensive mitigation and resilience. Behavioral denial is an extension of certain types of protection and offensive operations. A belief that the defender is or could utilize random patrols around a potential target, targeted killings and/or proactive arrests or detentions increases an adversary's uncertainty about the ability to effectively marshal and channel its available resources into a successful attack with minimal costs. In contrast, behavioral delegitimization is much more abstract in nature: it is primarily defined by discursive behaviors rather than concrete ones.

A key point is that denial communication rarely happens in a vacuum. Indeed, denial is often more of an afterthought or at least a secondary purpose to some other attempt to either protect valued assets or reduce the adversary's capabilities. Unfortunately, this multiplicity of strategic intent makes operationalizing and testing the effects and effectiveness of dissuasion by denial particularly difficult. Brief illustrative examples of dissuasion by denial from Israel and Northern Ireland are provided in the conclusion as a way to spur further research on the topic and identify next steps in the study of coercion, deterrence, and denial in counterterrorism.

EVALUATING DISSUASION BY DENIAL EMPIRICALLY

Although dissuasion by denial can include a wide range of behaviors, empirical focus here is placed on one of the most basic interventions: the building of dividing walls. Dividing walls are located far enough from the protected target that they are unlikely to be purely defensive. Likewise, while they can be enlarged and reinforced, dividing walls are essentially static fixtures in the strategic landscape. As such, their effectiveness in protecting a given target can relatively easily be assayed to alter the adversary's perceptions of the status quo and are largely irrelevant for the logic of bundling. Given the non-offensive, non-bundling nature of these barriers, the main challenge is differentiating between the protective and dissuasive effects of these barriers in order to evaluate dissuasion by denial. Protection might be assessed by answering two questions: Did the barriers stop or repel attempted attacks by the adversary? Was the protected target still attacked? Dissuasion might be assessed by answering a further four questions: Did the rate of attempted attacks decline after the wall was erected? Did the adversary adopt costlier tactics to overcome the barrier? Did the adversary adopt longer, riskier routes to reach the target? Did the adversary shift to attack different targets? A cursory exploration of the Israeli Security Fence and the Northern Irish Peace Line, despite their different intent—the former was meant to prevent suicide bombers from reaching targets in Israel, the latter was initially intended to stop mobs moving in both directions (mostly attacking Catholic enclaves) and gradually developed to stop a variety of different kinds of attacks (mostly emanating from Catholic enclaves) —provides some preliminary insight into assessing dissuasion by denial.

The main portion of the Israeli Security Fence was completed in August 2003 and strongly correlates with a dramatic reduction in the number of suicide attacks: fifty-three attacks in 2002, twenty-six in 2003, twelve in 2004, eight in 2005, six in 2006, one in both 2007 and 2008, and none in both 2009 and 2010.[31] Although this significant decline is likely the result of a combination of numerous factors, including a devastating Israeli offensive campaign targeting their adversaries' operational leadership, both protective and dissuasive influences of these barriers on the adversary's behavior are evident. Studies have identified clear patterns of bombers avoiding the barriers by targeting or traversing areas where construction had not yet been completed, choosing less direct, more circuitous routes around the barriers and coopting Israeli-Arabs to undermine the effectiveness of Israeli checkpoints at these barriers.[32] In addition, there has been a significant tactical shift from suicide bombings to stand-off weapon attacks, a topic explored in depth in Adamsky's chapter and in passing in Wilner's chapter.[33] The increased costs and risks, and reduced benefits these workarounds entailed greatly reduced the effectiveness of suicide bombings, but they did not eliminate the desire to carry out these attacks: in 2004, twelve out of 171 attempted suicide bombings (7%) were successful; in 2005, eight out of fifty-four (14.8%) were successful; in 2006, six of forty-eight (12.5%) were successful; in 2007, one of forty-four (2.3%) were successful; in 2008, one of sixty-four (1.6%) were successful; and in 2009, none of thirty-six (0%) were successful.[34] This pattern indicates that some sort of cost perception-adjustment process was in effect in the mid-2000s. After a terrible performance in 2004, the year after the first fence was completed, the adversaries drastically curtailed the number of attempts to achieve nearly the same number of successful attacks. It is likely that they concentrated additional resources into selecting and operationalizing these attempts in order to overcome the higher cost of penetrating the security fences.[35] However, it is interesting that this pattern was not repeated in the late 2000s: the number of attempts was largely flat from 2007 on, but the success rate became increasingly abysmal. Thus, the Security Fence clearly had a significant protective effect by stopping many attacks. It also appears to have had a strong but incomplete dissuasive effect: it significantly reduced adversaries' ability to reach their most desired targets with suicide bombers (an increase in operational cost), resulting in target-shifting to more vulnerable populations not protected by the fence and tactic-shifting to increased use of rocket attacks, but it did not completely dissuade the Palestinian groups from attempting suicide bombings, which remain an important part of their repertoire.

Moving to Northern Ireland, the first portions of the Peace Line in Belfast were erected in 1969 in the wake of large sectarian riots in Belfast; the British Army was deployed to restore order, manning positions between the restive neighborhoods.[36] Thus, the initial purpose was to temporarily separate the clashing communities. As the conflict changed, the Peace Line evolved to serve additional purposes, including preventing stones and bombs from being thrown into opposing communities, and blocking escape routes used by paramilitary groups after an attack.[37] By 2005, a mapping survey identified forty-one distinct barriers in Belfast that had been constructed by the Northern Ireland Office (NIO), and a large but indeterminate number of barriers erected by other agencies "which also serve to divide, separate and protect but which are not formally recognised as 'security barriers'."[38] A 2011 report, which adopted a more inclusive definition of Peace Lines, identified 99 barriers in thirteen distinct clusters.[39] Given the mixed intent behind their erection, it is unsurprising that the Peace Line has a mixed record of success. In terms of preventing sectarian mobs from invading and attempting to ethnically cleanse neighborhoods, the walls appear to have been a marked success; the threat of violent mobs overrunning vulnerable enclaves effectively ceased in 1969. As a counterterrorism tool, however, the barriers appear to have had a more limited effect on the paramilitaries' behaviors. Republican paramilitaries frequently targeted the security patrols that entered Catholic-dominated areas, so these targets of convenience reduced the need to travel to Protestant-dominated areas even before the barriers made this more difficult. Thus, the Peace Line had little relevance for a large portion of Republican attacks. Nevertheless, the Republicans continued to successfully attack targets in these more-difficult-to-reach areas throughout the conflict, often using innovative engineering solutions to pass through checkpoints undetected or to bypass defenses altogether, by using mortar bombs for instance. The paramilitaries added complexity and risk to their bombing efforts by smuggling the weapons piecemeal and then assembling them on the far side of the barrier. In addition, they shifted their tactics away from bombs to smaller, more easily concealed incendiaries, sniping and hoaxes.[40] On the other side, Loyalist paramilitaries often had connections in the security forces that made the barriers more permeable in reality than on paper. However, overuse of these resources, especially to carry out high-profile attacks, risked exposure of the asset in the security forces. Therefore, crossing through the Peace Line or the checkpoints necessary to circumvent those barriers likely did impose some cost for both types of paramilitaries.

Although the amount of paramilitary violence generally declined or shifted out of Belfast as more and better walls went up, it is difficult to assign a high degree of causality given all the other changes in the strategic environment. However, two studies of the geospatial distribution of attacks in Belfast provide tantalizing indicators that the Peace Line may have been a significant piece of the counterterrorism puzzle.[41] When violence is aggregated to the Census enumeration district, both studies found that most of the violence was concentrated in the economically deprived ghettos closest to the Peace Line barriers. Unfortunately, neither study differentiated between types of barriers nor assessed more granular measures of proximity to the barriers that could capture changes in routes. Moreover, they both used a pooled regression technique that aggregated seventeen years' worth of violence (1980–1997), which masks the potential temporal effects of the roughly two dozen barriers built during this time period; the Peace Lines were generally built where violence was the most prevalent. Nevertheless, the victimization patterns identified in these two studies suggest that the barriers may have funneled violence towards the portals within the Peace Line. Whereas the Peace Lines certainly made major cross-barrier attacks more difficult, they did not stop them entirely; the very limited data about attempted attacks foiled by these barriers makes it difficult to assess the full extent of their protective effect. However, the apparent shifts to more frequently targeting locations outside the cordons and to more costly and sophisticated engineering solutions or less impactful tactics to overcome the protective effects of the Peace Line indicate these barriers had a significant dissuasive effect.

CONCLUDING THOUGHTS: NEXT STEPS FOR THEORY AND EMPIRICISM

The theoretical discussion of the logics of counterterrorism provided at the beginning of this chapter highlights the internal inconsistencies in current applications of deterrence by denial in counterterrorism. In response, this chapter proposes a more logically coherent and exclusive version of the concept dissuasion by denial, which can be usefully applied in countering terrorism and militancy. Although more stringent about the type of influence being attempted, dissuasion by denial still encompasses a fairly wide range of behaviors beyond the most obvious example of target hardening. As such, there is a wide menu of choices for empirically evaluating a core set of research questions about dissuasive effects of counterterrorism policies. The empirical highlights provided by the Israeli and Northern Ireland examples demonstrates the feasibility of this research program. Both cases illustrate the dissuasive effects of dividing walls but also underline the methodological difficulty of measuring these effects.

Despite these analytical limitations, the two illustrative examples highlight several conditions that may limit the utility of dividing walls for protection and dissuasion: presence of highly desirable targets on the far side of the dividing wall, proximity of desirable targets to the wall (i.e., reachable by weapons launched over the wall), amorphous or multi-faceted motivations for building the walls, and the degree of permeability of the barriers. Moreover, lessons from these two cases also highlight the amount of research that remains to be done. Although there has been some preliminary work to geospatially examine the effects of these barriers on adversary behaviors, much more analysis is needed. Evaluation of dividing walls' relative defensive and dissuasive effects would greatly benefit from a finer granularity in the spatiotemporal data on both attacks and attempts, as well as better information on the routes which adversaries took before and after barriers were erected.

Notes

1. Crelinsten, "Perspectives on Counterterrorism."
2. Early work modeled this interaction as fully rational two-player opposi-
    tional games. Lake, "Rational Extremism," 15–29; Bapat, "State Bargain-
    ing," 213–230. However, more recent work has provided a more nuanced
    understanding that groups are boundedly-rational in their simultaneous
    pursuit of multiple objectives. Abrahms, "What Terrorists Really Want,"
    78–105; Sri Bhashyam and Montibeller, "In the Opponent's Shoes," 666–
    680.
3. Enders and Sandler, *The Political Economy of Terrorism*; Siqueira, "Polit-
    ical and Militant Wings," 218–236; Bueno de Mesquita, "The Quality of
    Terror," 515–530; Ganor, *The Counter-Terrorism Puzzle*; Arce and San-
    dler, "Counterterrorism," 183–200; Rosendorff and Sandler, "Too Much
    of a Good Thing?", 657–671; Sri Bhashyam and Montibeller, "In the
    Opponent's Shoes," 666–680.
4. Horowitz, "Nonstate Actors and the Diffusion of Innovations," 33–64.
5. This taxonomy is a simplification of the 2011 Department of Homeland
    Security National Preparedness Goal's five missions: Prevention,
    Protection, Mitigation, Response, and Recovery, https://www.fema.gov/
    pdf/prepared/npg.pdf.
6. Sandler and Siqueira, "Global Terrorism," 1370–1387.
7. Johnson, Mueller, and Taft, *Conventional Coercion*.
8. Knopf, "The Fourth Wave in Deterrence Research," 1–33; Wilner, "Deter-
    ring the Undeterrable," 3–37; Long and Wilner, "Deterring an 'Army
    Whose Men Love Death.'"
9. Tetlock and Levi, "Attribution Bias," 68–88; Blakemore and Decety,
    "From the Perception of Action," 561–567; Brewer and Lambert, "The
    Theory-Ladenness," 176–186; Gutsell and Inzlicht, "Empathy Con-
    strained," 841–845.
10. Cragin and Gerwehr, *Dissuading Terror*.
11. Davis and Jenkins, *Deterrence and Influence in Counterterrorism*.
12. Enders, Sandler, and Cauley, "Assessing the Impact," 1–18.
13. Davis and Jenkins, *Deterrence and Influence in Counterterrorism*.
14. Lutes and Bunn, "Dissuasion and the War on Terror," 73–84.

15. Bundling builds on the work of Huth and Russett, "Testing Deterrence Theory," 466–501 and Wilner, "Deterring the Undeterrable," which highlight the conceptual similarity between deterrence and inducement.

16. Davis, "Deterring and Otherwise Influencing Violent Extremist Organizations (VEOs)."

17. Brimmer and Hamilton, "Introduction: Five Dimensions of Homeland and International Security," 10.

18. Snyder, "Deterrence and Defense" (Reprint), 129; Davis and Jenkins, *Deterrence and Influence in Counterterrorism*; Anthony, *Deterrence and the 9-11 Terrorists*; Geipel, "Urban Terrorists," 439–467; Knopf, "The Fourth Wave in Deterrence Research."

19. *Presidential Decision Directive 39*, 1.

20. Schelling, *Arms and Influence.*

21. See chapter 3.

22. Gibbs, *Crime, Punishment, and Deterrence*; Erickson and Gibbs, "Specific Versus General Properties," 390–397; Stafford and Warr, "A Reconceptualization," 123–135.

23. Blumstein, Cohen, Roth, and Visher, *Criminal Careers and "Career Criminals."*

24. Post, "Terrorist Psycho-Logic."

25. Wilner, "Deterring the Undeterrable."

26. Trager and Zagorcheva. "Deterring Terrorism," 87–123.

27. Smith and Talbot, "Terrorism and Deterrence by Denial," 16–17. See also Smith, "Strategic Analysis, WMD Terrorism."

28. Kroenig and Pavel, "How to Deter Terrorism," 21–36.

29. Dutter and Seliktar, "To Martyr or Not," 429–443, 601; Knopf, "The Fourth Wave in Deterrence Research."

30. Wilner, "Deterring the Undeterrable."

31. "2010 Annual Summary – Data and Trends in Terrorism," http://www.jewishvirtuallibrary.org/jsource/Terrorism/2010Review.pdf.

32. Palti, "Israel's Security Fence"; Kaplan, Mintz, Mishal, and Samban, "What Happened to Suicide," 225–235; Kliot and Charney, "The Geography of Suicide Terrorism in Israel," 353–373; Jackson, et al. *Breaching the Fortress Wall.*

33. See chapters 2 and 7; Hillel Frisch, "Motivation or Capabilities?", 843–869.

34. Schachter, "Unusually Quiet," 19–27.

35. Frisch, "Motivation or Capabilities?"

36. McKittrick and McVea, *Making Sense of the Troubles*, 55–56.

37. Ravenscroft, "The Meaning of the Peacelines of Belfast," 213–221.
38. Jarman, *BIP Interface Mapping Project.*
39. "Interfaces Map and Database - Overview," http://www. belfastinterfaceproject.org/interfaces-map-and-database-overview.
40. Jackson, et al., *Breaching the Fortress Wall.*
41. Mesev, et al., "Measuring and Mapping Conflict-Related Deaths," 83–101; Mesev, et al., "The Geography of Conflict and Death," 893–903.

CHAPTER 5

DETERRENCE AS STRATEGY

THE STRATEGIC IMPORTANCE
OF DETERRENCE BY DENIAL

James J. Wirtz

The US government embraces deterrence as the cornerstone of its national security strategy. The reason why deterrence is its preferred strategy is also clear. US politicians and citizens alike would rather deter the outbreak of war or the occurrence some unwanted fait accompli, rather than engage in conflict to deny an adversary its objectives. The fact that Washington possesses a superior military capability vis-à-vis likely opponents only bolsters this strategic preference. Possession of a global surveillance and precision-strike complex,[1] a wide range of far-reaching conventional power projection capabilities, allied partners and bases, and a robust nuclear capability provides the United States with a wide array of options to make good on deterrent or compellent threats. Given this vast and highly effective military capability, the use of threats to deter or coerce likely opponents readily suggests itself as a cost-effective way to achieve US national security objectives. Americans are the world's leading believers in, and practitioners of, deterrence. They are also more willing to adopt compellence as a strategy than they readily admit.

Since the end of the Cold War, however, US deterrent and coercive threats have yielded a limited record of success. Iraq invaded Kuwait, despite the obvious risk of US retaliation, and then failed to yield to coercive pressure to abandon its ill-gotten gains. Over a decade later, the Ba'athist regime was destroyed by a US invasion, but not before Saddam Hussein failed to succumb to US coercion by complying with a list of demands to come clean about Iraq's weapons of mass destruction programs. It also appears that coercive threats have done little to slow the development of nuclear weapons by North Korea. The United States was attacked by al Qaeda on September 11, 2001, which is an act that calls into question the basic tenants of deterrence itself. Observers might counter that the failure of deterrence or coercion surrounding these events was idiosyncratic or tied to a failure to make threats clear to the opponent. Yet, it hard to believe that those who drove an airplane into the Pentagon, the headquarters of the US military, actually expected no response on the part of the US government to such a direct affront. The possibility exists that something is amiss in US deterrent strategy.

To identify this flaw in the concepts and practice of US deterrence strategy, the chapter first provides a typology of defense strategies and characterizes the dominant conception of defense by deterrence incorporated into US National Security Strategy. It then describes what is meant by conceiving of deterrence as a "strategy" and why opponents believe they can defeat this strategy without suffering consequences outlined by extant deterrent threats. It then goes on to reconceptualize the strategy of deterrence by denial to overcome the flaws in existing deterrence strategies that create a situation that can be exploited by potential opponents. The argument rests not so much on a critique of deterrence theory but on the notion that a flawed strategy for the use of deterrence reduces the credibility of deterrent threats. In other words, opponents believe that they can defeat US strategy, which makes execution of deterrent threats irrelevant to ongoing events.

A TYPOLOGY OF DEFENSE

Defense takes two forms: denial or deterrence. As figure 2 shows, however, this simple distinction is a bit complicated. Denial denotes the intention to prevent opponents from achieving their objectives by undertaking defensive measures at the point of attack or generally defeating some sort of military action. Instead of threatening retaliation after some unwanted activity or fait accompli, defense via denial implies meeting an opponent's military attack with a countervailing blow that defeats the initiative. Denial can also imply a willingness to undertake a preemptive attack or a preventive war to stop an opponent from undertaking an undesirable activity. As a strategy, undertaking defense via denial implies that the defending party is prepared to defeat the opponent regardless of the circumstances of attack. It also implies that defense is not based on the expectation that the opponent will moderate their behavior. Deterrent threats might be part of a defense by denial strategy, but the strategy itself does not rest on deterring war, it rests on the ability to defeat the opponent to deny them the ability to achieve their objectives. Defense by denial is the most commonly understood meaning of the term "defense."

Figure 2. Typology of Defense.

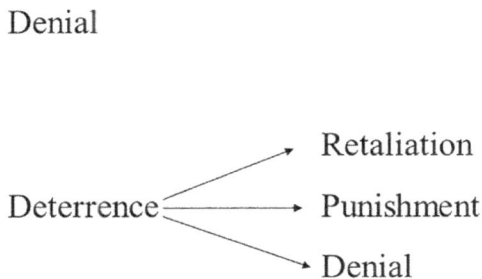

Denial

Deterrence → Retaliation
Deterrence → Punishment
Deterrence → Denial

By contrast, deterrence comes in three varieties: punishment, retaliation, and denial.[2] All three varieties of deterrence are based on the possession of the political, economic, military, and diplomatic capability to execute threatened action. Deterrence also rests on creating the belief in the opponent's mind that the threatened action will actually take place if "red lines" are crossed, which is often described as the "credibility" of the threat. Some scholars and policy makers believe that red lines should be clearly stated, so that opponents do not inadvertently cross them leading to failures of deterrence that might have otherwise been avoided.[3] Others believe that more ambiguity is called for in the making of deterrent or coercive threats in order to broaden their potential effect or to prevent opponents from adopting "salami tactics" (i.e., the use of incremental initiatives to gradually achieve objectives without crossing clearly delineated red lines).[4] Capability (the material capacity to make good on a threat should deterrence fail) plus credibility (the belief in the mind of the opponent that the party making a threat will actually be able and willing to execute the threat) are generally considered to be key elements of any deterrent or coercive strategy.

Deterrence or coercion by punishment is based on the promise that the costs of some unwanted activity on the part of the opponent will exceed the gains secured by achieving that objective. The costs—inflicted by sanctions, diplomatic action, or a military strike—do not have to be directed at reversing the unwanted gain, but they are instead inflicted by holding some valued asset at risk. For instance, to deter a territorial incursion, one might threaten to hold military forces or bases, urban centers, or industrial infrastructure at risk. In other words, retaliation would not be based on military action to directly reverse the territorial incursion but rather to inflict "unacceptable" costs on the opponent for their land grab. If the side that adopts a deterrence strategy based on punishment has the capability and will to make good on their threatened response and can clearly communicate their intention to retaliate prior to some unwanted action, then the prospect of suffering "unacceptable" costs should, ceteris paribus, deter the occurrence of the unwanted military activity in the first place.

Deterrence or coercion by retaliation is based on the promise of inflicting costs continuously on an opponent until compliance with a coercive threat is achieved or the gains achieved by some unwanted activity are abandoned. Unlike deterrence by punishment, deterrence by retaliation implies that costs—inflicted by sanctions, diplomatic action, or military activity—are not "all or nothing" in nature but will persist, continue, or even escalate in the face of some unwanted initiative. In the aftermath of a territorial incursion, for example, one might threaten to conduct military strikes at regular intervals until the opponent abandons their ill-gotten gains and returns to the status quo ante bellum. Opponents thus have the opportunity to abandon their gains as the costs of their initiatives begin to grow. Deterrence or coercion by retaliation thus allows the party practicing deterrence to fine tune the execution of their threats over time, while providing them an opportunity to generate additional political, military, or economic capabilities to execute deterrent threats. By expanding the time horizon for the execution of deterrent threats, deterrence by retaliation also involves a running contest over issues of credibility. In other words, credibility might rest on the opponent's belief in the ability of those issuing deterrent or coercive threats to commit to the sustained infliction of costs over an extended period.

Deterrence or coercion by denial is based on the promise that a response to some unwanted act will directly prevent the opponent from achieving its objectives. In other words, deterrence by denial does not rest on the threat of inflicting unacceptable costs on the opponent but instead promises to prevent them from achieving undesirable objectives in the first place. To deter a territorial incursion, denial thus might involve (1) threats to launch a preemptive attack or preventive war to deny the opponent the military capability to launch an attack; (2) threats to defeat an attack where and when it occurs by fighting at the border or launching a counteroffensive to reverse some ill-gotten gain; or (3) threats to eliminate the "bone of contention," so to speak, before it can fall into the enemy's hands, leading to a situation where temporary success is transformed into a pyric victory not only for the opponent but perhaps even both sides in the contest. Although this third option might strike some observers as self-defeating or unrealistic, war sometimes creates enormous death and destruction. In other words, launching a preemptive attack or preventive war or the effort to stop an attack where it occurs could very easily bring about the destruction of the object of contention.[5]

Several observations can be made about this typology of defense. First, defense by denial must be considered a preferred strategy because its ultimate success does not rest in the hands of the opponent. Regardless of what the opponent does—salami tactics, limited attack, or massive attack—one is prepared to reverse the initiative; control of the situation is shifted, at least in theory, to the defender. By contrast, the outcome of a deterrent situation actually rests in the hands of the party that is to be deterred from some unwanted action—whether deterrence or coercion succeeds or fails rests on their response to threats. Critics of deterrence are quick to point out that this is a rather complicated and uncertain process. According to Keith Payne, "For a strategy of deterrence to work by design requires: attentive players; the expression of threat; mutual recognition; communication and understanding; purposeful decision-making by the target audience based on rational calculation of expected risks costs and benefits, and the decision to yield to the threat; and the implementation of that decision. The success of the strategy can break down in any point of that process."[6] In effect, it is best to secure one's security by taking matters into one's own hands and to not base one's security on the rationality, competence, or compliance of a potential opponent.

Second, defense employing a deterrence-by-punishment strategy is probably best suited to situations in which it is impossible to reverse or deny an opponent the gains that flow from some unwanted action. In this situation, it is best *ex ante* to alter in the perception in the mind of the opponent that the potential costs from some unwanted action clearly outweigh potential gains. This strategy came to characterize US nuclear doctrine during the Cold War. The United States lacked the capability to deny the Soviet Union the opportunity to destroy the United States, and nothing could be done to reverse catastrophic levels of destruction once they occurred. By threatening to retaliate following this sort of attack, however, the United States intended to inflict similar or even greater levels of destruction against the USSR, effectively nullifying any benefits the Soviets might have gained by destroying the United States by visiting catastrophic levels of destruction on the USSR.

Third, defense employing a deterrence-by-retaliation strategy is probably best suited where it is possible to reverse the gains enjoyed by an opponent that undertakes some undesirable action. Punishment relies on what Thomas Schelling called "the diplomacy of violence," by directly undertaking countervalue strikes without inflicting a strategic or tactical military defeat on the opponent first.[7] Deterrence by retaliation does not require a finely crafted denial strategy tailored to counter specific insults, just a general ability and will to inflict pain on an opponent following a fait accompli or attack. Retaliation thus holds out the promise of returning to the status quo ante bellum by undertaking military action suited to the defenders military capabilities, not to some specific action on a field of battle chosen by the attacker to suit its needs, strengths and strategies.

Fourth, defense that relies on deterrence by denial should probably be treated as a hybrid strategy that obviously and closely integrates the use of deterrence to prevent war with actual plans to fight and win the war if deterrence should fail. In that sense, it is a politically and strategically challenging strategy because it forces politicians and military strategists to assess and deconflict *ex ante* the warfighting strategies tailored to meet specific contingencies and the political and diplomatic demands created by the need to communicate deterrent or coercive threats to likely opponents. Put somewhat differently, politicians and military strategists must understand and accept what can or must be done in the aftermath of deterrence failure before they can credibly threaten a deterrence-by-denial strategy. Credibly communicating the ability to fight and win a conflict is especially challenging in the realm of conventional war because these types of threats are inherently "contestable," in the sense that opponents could hope to overcome denial threats by defeating the opposing military's operations. And, as history demonstrates, commonly accepted metrics for assessing the military balance *ex ante* are often woefully inadequate when it comes to predicting the outcome of conflict, which leaves a bit more than a glimmer of hope to underdogs that are threatened by seemingly more powerful opponents.[8]

Several qualifications also could be offered about this typology of defense. For example, all components of the typology are most applicable in symmetrical conflicts involving conventional weapons, or in settings involving conventional extended deterrence. Nevertheless, in theory and practice, certain preferred strategies emerge. Deterrence by retaliation is most often associated with nuclear weapons. Deterrence-by-denial strategies generally involve conventional weapons, although a few theorists suggest the possibility of adopting nuclear denial strategies.[9] Under some circumstances, for example, when facing nihilistic opponents, defense by denial would be a preferred option—deterrence is not a panacea and will not succeed if the opponent cannot be readily identified, or is extremely risk and cost acceptant, or does not possess targets of any value or significance that can be held at risk.

Several observations also can be drawn from this typology of defense when it comes to the US conception and practice of deterrence. The US approach to deterrence is a bit schizophrenic, especially when it comes to deterrent or coercive threats intended to preserve regional status quos. Deterrent or coercive threats *ex ante* are often stated in a general way, highlighting what behavior is deemed unacceptable without stating or even specifying which type of deterrence is actually being threatened. The ambiguity about exactly what might follow should deterrent or coercive threats fail highlights the simple fact that policymakers are hoping that deterrence or coercion will succeed. According to Brad

Roberts, "When confronted with the possibility that adversaries ... might actually doubt America's resolve to defend a stake they seek, it is common for Americans to respond with the convictions that "they wouldn't dare" and "we'll turn them into a glass parking lot (that is we'll attack them with nuclear weapons until there is nothing left of them). These are not theories of victory. They are forms of wishful thinking."[10] To plan for deterrence failure confronts US strategists with the immediate possibility that their overall plan to avoid war might be flawed or that coercive or deterrent threats might prove to be self-defeating, thereby exacerbating—not containing—the factors propelling both parties into a conflict. A cold assessment of what might happen in the wake of deterrence failure also might lead them to abandon deterrence and instead switch to a defense by denial strategy.

When it comes to executing deterrent threats, what unfolds is largely driven not by prior design but by an ad hoc US response to strategic or policy failure. In the wake of the al Qaeda attack on US embassies in Africa, the Bill Clinton administration, in an effort to inflict costs against the terrorist syndicate, executed punishment strikes against al Qaeda training camps in Afghanistan. In the aftermath of the al Qaeda attacks against the United States on September 11, 2001, however, the George W. Bush administration undertook a deterrence-by-denial campaign in an effort to destroy al Qaeda's ability to launch further attacks against the United States and its friends and interests overseas. In response to Slobodon Milosovich's campaign of ethnic cleansing, the United States not only used retaliation to alter Serbia's behavior but also made a half-hearted effort to launch a denial campaign to force Serbian units to abandon Kosovo. In over a decade of conflict with Iraq, the United States first launched a deterrence-by-denial campaign to eject Iraqi forces from Kuwait, which was followed up by a series of retaliatory strikes for Iraq's failure to comply with UN disarmament demands (Operation Desert Fox). The culminating round in the drama was a defense by denial campaign (i.e., preventive war) to finally rid the world of the Ba'athist regime. Until they were diverted by Russian diplomacy, the Obama administration was preparing to undertake punishment strikes against Syria to inflict costs upon the regime for its use of chemical weapons against its own people, preparations that appeared to be entirely ad hoc after Bashar al-Assad crossed the "red lines" that were alluded to by the president in April 2013.

DETERRENCE AS STRATEGY

As Colin Gray tells us, strategy is the idea that informs one party's actions when it comes to shaping the political choices of an opponent in a favorable way. It involves using all of the political, diplomatic, economic, and military resources at one's disposal in a purposive way to effect the opponent's behavior and outlook, while seeking to expand, if not restricting, one's own political or military ability to achieve one's objectives. Gray uses the metaphor of a bridge when it comes to describing strategy because strategy is the way one bridges the gap between capabilities and political objectives. Strategy also helps bridge the chasm created by the dialectical nature of conflict and war. In other words, the outcome of conflict is produced by the interaction of at least two parties, and strategy is a course of action that manipulates conflict's dialectic to one's advantage. Strategy is about constraining the opponent's conception of what is politically or militarily desirable or possible while not allowing our choices to be constrained by the opponent's strategy.[11]

Deterrence is a sophisticated example of the notion of strategy outlined by Gray. Deterrence is intended to influence an opponent's political and military preferences and outcomes, and deterrent relationships are determined by the interaction of both opponents. All deterrence strategies and all varieties of deterrence rely on the notion that by credibly threatening some sort of action, the opponent will choose not to fight, which in the end will create a situation in which the party making deterrent threats will not have to fight either.

Given this conception of strategy, it is relatively easy to discern the outlines and weaknesses of the deterrence strategy practiced by the

United States. US policymakers view deterrence as a war-prevention strategy that often fails to incorporate notions of what might transpire if deterrence fails, which is a fatal weakness. In other words, little thought is given in advance to determine whether deterrent threats are based on denial, punishment, or retaliation, and even less thought seems to be given to operational planning geared towards supporting general deterrence threats. The workings of this strategy are also clear. The strategy is based on highlighting, to friend and foe alike, that given US conventional force-projection capabilities and reconnaissance and surveillance assets, no person, military unit, organization, government, or nation is beyond the reach of the long arm of Washington. In other words, American policymakers and strategists seem to believe that US military capability speaks for itself and that an appreciation of this capability is enough to cow potential opponents into submission.[12] Because their mind comes to rest on the notion that the sole purpose of deterrence is to prevent the outbreak of war, little consideration is given to planning for the failure of deterrence or if the opponent discerns a way to overcome deterrent threats.

Despite the fact that the United States enjoys truly overwhelming military superiority over most likely opponents, US deterrent threats lack credibility because opponents are coming to understand that US deterrence strategy itself is not credible. In other words, no one doubts the US physical capability to execute a wide gamut of deterrence threats, but considerable questions can be raised about US willingness to execute threats should the need arise.[13] US deterrence strategy is a war-prevention strategy geared towards peacetime and maintaining the peace, but it leaves open the question of what will happen should deterrence fail. Opponent's do not doubt US capability to act on threats, what they are banking on is a lack of US will to act on those threats in the aftermath of some insult to the status quo. In other words, they are devising ways to inflict a strategic defeat on the United States by causing its "war-prevention" general deterrence strategy to fail at the outset of hostilities, thereby forcing Washington to confront a fundamental political strategic assessment at the most inopportune time.

Three sources of optimism often animate thinking when challengers consider US deterrent strategies.[14] First, leaders sometimes believe that they can capitalize on strategic surprise to circumvent deterrent threats, presenting opponents with a fait accompli that cannot be easily overturned. Strategic surprise and the failure of deterrence relationships are clearly linked in the history of international crises and the outbreak of war. The party to be deterred becomes captivated by the possibilities created by some surprise military initiative, which in their minds will allow them to present the party issuing deterrent threats with a fait accompli that will effectively nullify a deterrent threat. A fait accompli makes both conventional and nuclear deterrence irrelevant because it presents the party banking on deterrence with strategic failure before it can bring its superior force to bear. Deterrence failure thus transforms the conflict into a test of who is willing to engage in an attritional struggle to reverse the status quo. The party challenging the deterrent threat is in fact banking on the fact the party relying on deterrence will fail that test.

Second, the targets of deterrence sometimes believe that they can capitalize on an international political setting or on the domestic politics of opposing powers that will prevent them from actually executing a deterrent threat or, if executed, will prevent the them from bringing the full force of their military power to bear against them. As the North Vietnamese began to justify their decision to launch the Tet Offensive, for instance, General Vo Nguyen Giap offered an explanation of why the United States would not be able to bring its superior power to bear to stymie Hanoi's military effort to unify Vietnam:

> The US imperialists must cope with the national liberation movements [in countries other than South Vietnam], with the socialist bloc, with the American people, and with other imperialist countries. The US imperialism cannot mobilize all their forces for the war of aggression in Vietnam.[15]

Giap recognized that the United States possessed the military resources needed to end the conflict quickly, but he also believed that it faced competing interests and pressures that would restrain its freedom of action in Vietnam. Similarly, Saddam Hussein believed that "casualty aversion," a domestic political constraint, would prevent the United States from interfering with his occupation of Kuwait. As the United States and an international coalition increasingly appeared to use force to eject Iraqi forces from Kuwait, Hussein apparently believed that Soviet opposition to American intervention would deter US military action in the Middle East.[16] When Moscow, preoccupied with the collapse of its empire, failed to protect its client, Hussein berated the Soviet leadership for failing to act like a Superpower:

> He who represents the Soviet Union must remember that worries and suspicions about the superpower status assumed by the Soviet Union have been crossing the minds of all politicians in the world for some time.... Those concerned must choose this critical time and this critical case in order to restore to the Soviet Union its status through adopting a position that is in harmony with all that is just and fair.[17]

Both the North Vietnamese and the Ba'athist regime in Baghdad believed that the threat posed by other Great Powers, international political opposition, or domestic political restraints, would be sufficient to hold superior US military power at bay. In other words, deterrence failure would alter the strategic setting, bringing potent international or domestic political forces into play, making it less desirable to execute deterrent threats.

Third, challengers can come to believe that moral or political constraints, arising from international or domestic public opinion that emerges in the course of some provocation will restrain the strong, especially if threatened retaliation seems out of proportion or misdirected against innocent bystanders. In other words, opponents can come to believe that eruption of violence itself will force the party issuing deterrent threats to reassess the utility of the use of force in general, or the execution of specific deterrent threats. Concerns about the manipulation of material conditions to directly alter political perceptions regarding the relevance and effectiveness of traditional military options is a reoccurring concern among military analysts, particularly in recent decades. "Fourth generation warfare" is one of the latest terms used to describe the effort to influence outcomes in war through political, not military instruments.[18] According to Thomas X. Hammes, "Fourth generation war uses all available networks—political, economic, social and military— to convince the enemy's political decision makers, that their strategic goals are either unachievable or too costly for the perceived benefit. It is rooted in the fundamental precept that superior political will, when properly employed, can defeat greater economic and military power."[19] Fourth-generation warriors are not focused on defeating superior military opponents on some battlefield. Instead, they focus on using civil disorder or low-intensity warfare to manipulate social, political, and cultural ties to alter local and global political perceptions in their favor. Innovative campaigns are designed to manipulate political perceptions of what is at stake in a given conflict and to create situations that make conventional military operations appear irrelevant or of limited utility. The goal is to create a situation in which the use of superior firepower and intelligence-surveillance-reconnaissance capabilities offer few good remedies to local turmoil. Under these circumstances, execution of advertised deterrent threats might be viewed by friend and foe alike as doing little more than exacerbating conflicts and increasing suffering among local civilians or third-party bystanders.

The key concept behind all these counter-deterrence strategies is clear. Destroy the deterrence strategy adopted by the United States, and the possibility is created to avoid the execution of deterrence threats. The critical weakness of US deterrence strategy flows from the assumption that given overwhelming US military superiority, it is inherently credible; thus, it cannot fail. Nevertheless, if an opposing party is highly risk acceptant, or motivated by interests more vital to it than to the United

States, shifting the strategic setting can defeat US strategy. Because US deterrence threats are made to prevent war and are not a blueprint for a war-winning strategy, US policymakers confront a series of stark choices in the aftermath of deterrence failure, choices that receive little consideration in peacetime. If they can defeat US deterrent strategy, opponents might create a situation in which US political and military options become highly constrained.

DETERRENCE BY DENIAL

If US deterrence strategy has been put at risk by the superior strategy of its opponents, what can it do to regain strategic supremacy? Here, a broad definition of deterrence by denial comes into play, a definition that entails the effort to defeat the opponent's counter-deterrence strategy. In other words, if opponents are determined to make deterrent threats irrelevant by taking actions that deliberately lead to deterrence failure, US policymakers must take actions to strengthen their deterrence threats so they are immune to this type of manipulation by opponents. Despite its overwhelming military superiority, the United States must take action to increase the credibility of its deterrence threats.

The notion that US policymakers must take action to strengthen the credibility of their deterrent and coercive threats would not come readily to their minds because they believe that US military superiority makes their threats inherently credible. Because they believe that they possess an "assured retaliatory capability" across the entire spectrum of conventional and nuclear options, no state, or non-state actor for that matter, would be willing to cross stated "red lines" and risk execution of US deterrent or coercive threats. They tend to believe that credibility flows directly from their superior capabilities. In a sense, opponents share their perception of US military superiority, but they believe that by defeating the US war-prevention strategy, they can actually succeed in achieving their objectives if hostilities actually erupt. They believe that by altering the strategic situation, they can cause US policymakers to reassess the desirability of actually making good on deterrent threats, and that this reassessment will actually work in their favor.

To preserve their war-prevention strategy of deterrence, US policy-makers must make a concerted effort to defeat the opponents' counter-deterrence strategies by taking deliberate efforts *ex ante* to prevent likely opponents from believing that the outbreak of war would actually be to their advantage by altering US cost-benefit calculations related to actual engagement in hostilities. By denying opponents the ability to execute their counter-deterrence strategy, the United States would bolster the credibility of its deterrent threats by strengthening its ability *ex ante* to actually execute its threats if deterrence or coercion fails. To meet the three counter-deterrence strategies available by opponents, the United States should contemplate three different types of initiatives.

The surprise attack counter-deterrence strategy offers opponents a prompt way of defeating US war-prevention strategies. By presenting the United States with a fait accompli, it alters the strategic setting by presenting US policymakers with a stark issue. Although cost-benefit calculations support a US deterrence policy *ex ante*, will cost-benefit calculations actually favor engagement in hostilities to reverse the fait accompli? Opponents are betting that the brave talk emanating from Washington *ex ante*, will not be matched by a willingness to pay the price in blood and treasure to reverse their ill-gotten gains.

The best way to defeat a surprise attack counter-deterrence strategy is by denying opponents the time and opportunity to devise some sort of exquisite military evolution to deliver a prompt setback to US interests. To do this, deterrence-by-denial strategies have to remain dynamic to confront opponents with an ever-changing military problem. This is not easy or inexpensive. Force structure, basing modes, patterns of operations, and command-and-control procedures must continually evolve to prevent the opponent from leisurely devising stratagems or operations that can exploit weaknesses that are inherent in any military posture. The pace of these changes should be dictated by the perception of risk generated by an opponent's incentives to alter the existing status quo. By definition, surprise attack is an extraordinarily risky enterprise that succeeds or fails by the finest of margins and incorporates a series of planning assumptions that must be valid if surprise is to achieve its military or political objectives. A dynamic defense-by-denial strategy complicates the opponent's planning environment, reducing confidence that the circumstances needed for a planned surprise attack will succeed will actually be present when the attack is actually executed. To defeat the surprise-attack counter-deterrence strategy, it is probably best to create some ambiguity in the mind of the opponent about the true state of one's defense by denial and defense by deterrence capabilities.

The second counter-deterrence strategy that could be embraced by opponents is to capitalize on an international or domestic political setting that will prevent the execution of a deterrent threat or, if executed, will prevent the occurrence of significant military action. To defeat this counter-deterrence strategy, the United States must increase its political ability to act on deterrent or coercive threats by building support *ex ante* for its specific political-military objectives and the premise that it is acceptable not just to threaten to use force to preserve the status quo but also to actually use force should deterrence or coercion fails. Currently, few efforts are explicitly undertaken to generate domestic or international political support for the implicit objectives of US coercive or deterrent threats—the lack of political support for these objectives only gains salience in Washington following the failure of threats to deter war or alter some unwanted behavior. Political, not military, constraints on US behavior are the Achilles' heel of US defense-by-deterrence strategies because opponents are banking that these constraints will emerge following the eruption of hostilities, forcing US policymakers to reassess their freedom of action to execute threats. By laying the political groundwork to support preservation of the status quo, policymakers are actually destroying the basis for this important counter-deterrence strategy, thereby increasing the credibility of their threats.

The third counter-deterrence strategy that might be embraced by opponents is to force the party issuing deterrent threats to reassess the utility of the use of force in general, or the execution of specific deterrent threats. To undermine this counter-deterrence strategy, the United States needs to broaden its political and military capabilities to equip itself with a broad range of options to prevent opponents from using social and political turmoil to achieve their objectives. This too is a demanding set of requirements because it forces Washington to devise political acceptable and militarily feasible responses to political turmoil, civil disorder, or general chaos that is not easily countered by precision-strikes launched at stand-off ranges. Here too there is an *ex ante* need to foster international and domestic political support for the notion of measured intervention or sustained international pressure to prevent opponents for capitalizing on mayhem among civilian populations to achieve their objectives.

Conclusion

Effective defense by deterrence is not based on capability alone. It is based on creating the perception in the mind of the opponent that threats will actually be acted on in the event the failure of deterrence or coercion. US opponents, however, are coming to understand that the United States has adopted a war-prevention strategy, and that this strategy can be defeated by the outbreak of war itself. If they initiate hostilities, they believe that they can confront Washington with a political and strategic failure, leaving to ponder its next move under circumstances that now favor the opponent's interests and objectives.

US policymakers can defeat these counter-deterrence strategies and preserve the peace. The first step in this process, however, is understanding that deterrence is a strategic relationship and that opponents will not be cowed by the superior military capabilities possessed by the United States. Threats to use force will appear credible in peacetime only if opponents believe that the United States will actually act on those threats should deterrence or coercion fail. Bolstering the credibility of

US deterrence threats has little to do with ample US military capability. Instead, it has everything to do with creating the political freedom to act as advertised if deterrence or coercive threats should fail.

There is no surefire way to create political support to honor deterrent or coercive threats and to communicate that impression to the opponent. Creation of a politically compelling justification for deterrence or coercion and integrating it into a defense strategy that appeals to domestic and allied audiences would be a good place to start; admittedly, this is advice that is easily given and difficult to put into practice. Beyond this obvious recommendation, however, one might suggest that deterrent and coercive threats should be crafted so that they literally "fail deadly," creating a political backlash that would support the execution of threats. Additionally, anything that reduces freedom of maneuver in the wake of the failure of deterrence or coercion bolsters credibility; doing so *ex ante* actually is a demonstration of a strong political commitment to make good on strategies of deterrence or coercion. Policymakers also should be aware that it might be impossible to establish the credibility of certain deterrent or coercive threats given the low stakes that underpin some commitments. To be effective, strategy has to correspond to the interests and objectives at stake.

Ultimately, this chapter highlights the fact that deterrence is a strategy that is intended to shape the opponent's politics and is in turn limited or enhanced by the domestic and alliance politics of the state seeking to prevent the outbreak of war. US policymakers and officers, who are mesmerized by their superior military capabilities, not only have failed to think through the military implications of their deterrent threats, but they also have apparently failed to fully consider the political requirements of their policies. Admittedly, it is impossible to tell if the American public and allies would be willing to support the complete range of deterrence-by-denial strategies that would have to be constructed to strengthen the general deterrent threats issued by the United States to preserve various regional status quos. But by not addressing the domestic and alliance politics of deterrence *ex ante*, US policymakers are playing into the hands of their opponents who see great advantage in being able to pick the time and place for this sort of US strategic assessment. The time they will pick is when the US war-prevention strategy fails, and the place will not be of Washington's choosing.

NOTES

1. Some of these capabilities, for instance, Conventional Prompt Global Strike, the ability to destroy virtually any target anywhere in the world within one hour, might also be more of an aspiration that a reality. It is unclear, however, how closely current U.S. capabilities approximate this aspiration. See Acton, *Silver Bullet?*.
2. Rhodes, *Power and MADness*, 90–100.
3. Tertrais, "Drawing Red Lines Right," 7–24.
4. Secretary of State Dean Acheson's National Press Club Speech of January 12, 1950, is often cited as an example of the articulation of a clear red line that led an opponent to adopt salami tactics. Acheson defined the American defense perimeter as a line running through Japan, the Ryukyus and the Philippines, which seemed to leave Taiwan and South Korea to their own devices. When the North Koreans marched south in 1950, Acheson seemed responsible for encouraging aggression. See James Matray, "Dean Acheson's Press Club Speech Reexamined."
5. In 1960, for instance, Navy observers noted that the execution of the "day alert" portion of the U.S. strategic nuclear force against targets in the Soviet Union would create a situation in which fallout exceeded dangerous limits for Helsinki, Berlin, Budapest, Northern Japan and Seoul. Rosenberg, "U.S. Nuclear War Planning," 55.
6. Payne, *The Great American Gamble*, 18.
7. Schelling, *Arms and Influence*, 1–34.
8. The relationship between the "contestability" of costs and outcomes and deterrence is explained by Harknett, "State Preferences, Systemic Constraints," 51–53. By contrast, nuclear weapons effects are not commonly contestable. In other words, despite defensive preparations or the skill of opposing forces, if a one-megaton nuclear weapon is air burst over an urban area, approximately fifty percent of those living within a five-mile radius of the detonation will die promptly. The outcome of especially large-scale nuclear war is thus relatively easy to estimate in advance.
9. Mearsheimer, *Conventional Deterrence*, 14–15; Glaser, "Why do Strategists Disagree," 113–117.
10. Roberts, *The Case for U.S. Nuclear Weapons*, 104.
11. Gray, *The Strategy Bridge*.

12. Roberts notes that Pentagon planners tend to rely on U.S. conventional and nuclear superiority alone as a general deterrent. "This has the virtue of greatly simplifying the problem of U.S. deterrence planning. It also appeals to those who think that American military strength translates automatically into successful deterrence." Roberts, *The Case for U.S. Nuclear Weapons*, 104.
13. Chamberlain, *Cheap Threats*.
14. The remainder of this section draws on Wirtz, "Deterring the Weak."
15. Giap, "The Big Victory," 237.
16. Hope that the Soviet Union (Russia) would somehow constrain U.S. freedom of action also seemed to influence Saddam Hussein's behavior leading up to the Second Gulf War and Slobadon Milosevich's actions in Kosovo in 1999.
17. Saddam Hussein quoted in Freedman and Karsh, *The Gulf Conflict 1990–1991*, 164.
18. For a description of how the issues identified by Hammes are in fact a long-standing development in international affairs, see Wirtz," Politics with Guns," 47–51.
19. Hammes, "War Evolves into Fourth Generation," 42.

CHAPTER 6

DENYING NORTH KOREA

ACTIVE DEFENSE TECHNOLOGY AS DENIAL

Jonathan Trexel

In Northeast Asia, the threat from the Democratic People's Republic of Korea (DPRK) is formidable: it has large numbers of offensive ballistic missiles, it may possess missile-capable nuclear weapons, and it has a propensity for aberrant or provocative behavior (e.g., export of nuclear and missile technology, military violence against its neighbors).[1] The DPRK's large stockpile of missiles means that it can, in a matter of minutes, endanger military, civil, and economic targets in neighboring countries like Japan. The North Korean security posture and coercive power-based approach to diplomatic and economic relations today are undergirded by its ballistic missiles. North Korea may have as many as 1,000 ballistic missiles in the field on 200 reloadable launchers.[2] Coercion is simply pressuring an actor to do something it does not wish to do. For North Korea, coercive tactics such as provocations—including missile flight tests—and other malign behavior are used against Japan to weaken US-Japan allied cohesion; decouple Japanese support for US force presence and crisis staging; extract diplomatic or financial and trade concessions and postwar reparations;[3] attain relief from varied sanctions; undermine a multiparty negotiating posture in nuclear weapons talks in which Japan has participated; and influence Japanese intra-war decisions by showing risk tolerance, even unpredictability. Ballistic missile defense (BMD) can serve to counter coercion.[4] However, whereas DPRK possession of ballistic missiles casts a threatening shadow to coerce Japan and others including those below the threshold of armed conflict, Japanese development and possession of advanced BMD technology provides Japan a potential deterrence-by-denial offset.

The thesis in this chapter is that one can counter an adversary's coercion strategy by deterring the use or employment of their means of coercion. More specifically, North Korea's principal means of coercion has been ballistic missiles and their testing; Japan's counter-coercion strategy was to deter North Korean use and employment of its missiles through development and deployment of highly credible missile defense technologies, thereby significantly denying North Korea its perceived benefits and objectives. This chapter will briefly describe and analyze Japan's deterrence strategy against DPRK coercive behavior from a theoretic and practical view. It presents BMD technology as a unique case study of deterrence by denial against ballistic missile-based coercion below the threshold of armed conflict. Unlike other deterrence cases, the post–Cold War Japan-DPRK relationship considers deterrence of a nuclear-armed state by a non-nuclear-armed state; deterrence in the absence of offensive kinetic military cost-imposition capabilities; and, deterrence effectiveness under "general deterrence" conditions below the threshold of armed conflict.[5] The chapter begins with an overview of deterrence and missile defense theory. Next, the two principal actors are reviewed in context of the Japan-DPRK relationship. Lastly, the chapter further unpacks the deterrence-by-denial dynamics of Japan's BMD system with respect to DPRK behavior.[6] BMD is not Japan's only technology or means of deterrence, but it is its principal means of deterring coercion by ballistic missile threats.[7]

DETERRENCE, DENIAL, AND MISSILE DEFENSES

Deterrence is defined here as "preventive influence over an adversary's decision calculus" and can occur across the potential conflict continuum (i.e., peacetime, crisis, and conflict).[8] Political leaders can devise deterrence strategies using all of the instruments of power at their disposal. Further, deterrence is no longer limited by many scholars to a fear of cost imposition; it can be accomplished in other ways including denial as well as affecting an adversary's perceptions of restraint or not acting.[9] In an early work, Glenn Snyder characterized denial as influencing the probabilities of an adversary achieving his objectives by, for example, holding one's territory, and by limiting damage to one's forces and nation.[10] From this view, Japan's BMD therefore can be considered both pure defense, in the sense of simply working to save lives under ballistic missile attack, and deterrence by denial, or what Thomas Schelling called "deterrent defense." As technologically developed defenses, BMD is intended to sway North Korean leaders from deciding to use ballistic missiles at all, deciding to use them only in limited fashion, and from choosing to engage in other unwanted behavior that could somehow escalate.[11]

Deterrence by denial can be accomplished in many forms using high-technology capabilities or through carefully planned procedures and tactics. High-technology denial by a defender can include hardening, redundancy, operations security, warning and detection, active defenses, jamming, stealth, and mobility. Procedural or tactics-based forms of denial can include resiliency and recovery, reconstitution and replacement, robustness, consequence management, civil defenses, attrition, surprise and unpredictability, dispersion, camouflage, concealment, deception, posturing, and, when necessary, defensive attack operations. Ballistic missile defenses are considered an active defense technology that contributes to deterrence by defensively denying an adversary a host of potential benefits. Japan's BMD system is considered second in the world only to the US system in terms of technological quality and, as such, presents a good case for considering denial through active defense technology.

Further, for an assessment of this case, deterrence can be divided between *absolute* and *specific* versus *restrictive* deterrence. Absolute deterrence—what some call the classical model—exists when an actor continuously refrains from a particular egregious action, such as waging war or employing nuclear weapons, often due to credible threats of retaliation. Specific deterrence also deals with deterring a particular action but in this case is tailored to a specific actor or leader in a deterrence relationship. Oftentimes these two ideas are combined so the deterrence intent is to deter a specific actor from ever carrying out a specific deed in specific circumstances. This form of deterrence is binary; that is, it seeks to prevent an adversary behavior absolutely. If the adversary acts against our deterrent intent, that would be deterrence failure. Restrictive deterrence, conversely, is a form of deterrence applicable to certain types of behavior such as crime. In this case, society understands it cannot prevent all crime but seeks to manage it instead. Restrictive deterrence seeks favorable modifications in adversary behavior, such as broad reductions in the frequency, intensity, or scope of malign behavior. Restrictive deterrence is more applicable to Japan's suasion of DPRK's decision-making regarding coercion through its missile testing problem. Whereas even a single missile overflight is one too many, for Japan it would be unrealistic to deter every coercive missile launch over a span of several years where dozens of flight tests are expected to occur. Rather, in keeping with restrictive deterrence, Japan most likely seeks to use its BMD to limit the frequency, scope, and intensity of North Korea's ballistic missile flights, thus effectively denying the DPRK its coercive objectives sought through ballistic missiles or their testing. North Korea has proven immune to many penalties imposed by various actors over the years, continuing instead to test their missiles as a form of coercion despite the fears of such costs. However, mechanisms of deterrence by denial, such as active defense technologies like Japan's BMD, directly challenge DPRK calculations, not necessarily through traditional fear of punishment but by actions such as complicating their planning, precluding hoped-for benefits, undermining the value of their source of coercion, and obscuring a confident calculus of DPRK's future position.[12]

The principal purpose of missile defense is to provide crisis or wartime kinetic intercept and destruction of ballistic missiles, cruise missiles, and aircraft in flight. Most theorists and practitioners agree that missile defenses are an important relatively new military technology, though disagreement exists as to their efficacy as a tool of deterrence. In very broad policy terms, it is argued that missile defenses are used to dissuade the development of ballistic missiles, deter their use if developed, and defeat their use in attack.[13] Missile defenses provide opportunities for other benefits to nations under missile attack, including damage limitation to military forces, population centers, industry, leadership, and critical infrastructure. Importantly, the advantages that missile defenses provide the defender can also influence the attacker's decision-making by convincing him he will be denied the operational benefits perceived or sought—at least to the degree he believes BMD is fully effective— and denied strategic political benefits perceived by the attacker with the use of his missiles whether BMD is effective. BMD could also affect an adversary's decision-making regarding his efforts to reshape his position through day-to-day actions, as well as psychologically, such as thwarting adversary attempts to appear unpredictable with his use of missiles. Use of missile defense decreases the adversary leader's perception of the probability the missile attack will successfully reap perceived benefits as a consequence of a missile attack and illustrates but one way missile defense can deter an aggressor from initiating the attack or limiting the value of missile activities, such as missile tests, to coerce. One way that BMD achieves its denial effect in potential shoot-respond scenarios is by raising adversary leader uncertainties about the outcomes of his attack.[14]

FRAMING JAPAN'S DETERRENCE CHALLENGE

The regime in North Korea is today a threat to Japan with ballistic missiles, weapons of mass destruction (WMD), and provocative behavior.[15] North Korea's ballistic missile program includes large numbers, types, and ranges of missiles, some of them assessed to be capable of carrying WMD.[16] The most threatening missiles to Japan are the Nodong and Musudan. Such missiles are road-mobile systems making them very difficult, if not impossible, with current technologies to find and destroy before they were launched, allowing North Korea to use them in surprise attacks on Japan.[17] Tokyo, as the center for Japan's economic and political activity, is especially vulnerable to Nodong.[18] The Musudan and Nodong missiles could combine to threaten Japan with as many as 500 offensive ballistic missiles. For Japan, short of war the dominant threat stems from North Korean coercion by using the threat of limited ballistic missile attacks backed up by missile tests. More specifically, Sugio Takahashi states limited North Korean ballistic missile attacks or "cheap shot" raids on Japan fall below the "threshold for retaliation" by the United States under extended deterrence conditions.[19] Even repeated North Korean ballistic missile tests that threaten Japan's population with overflight are considered "coercive diplomacy," implicitly or explicitly threatening punishment for noncompliance with some DPRK political demand. Such sub-conflict behavior by North Korea toward Japan places the onus for the defense—and deterrence—of such threats upon Japan.[20]

As Japan's security evolved in the post–Cold War period, it was faced with three overall policy choices: rely wholly upon the US nuclear umbrella to manage threats; depend increasingly upon its own conventional capabilities, including BMD technology for deterrence and defense; or develop its own offensive nuclear weapons capability.[21] In 1998 North Korea tested a Taepodong-1 long-range ballistic missile that flew over Japan, an event that frightened Japan's population and surprised its leadership. Following this test, Japan's approach was to address the threat militarily through deterrence, BMD, and other activities.[22] This event stirred Japan's leaders toward BMD technologies for three reasons. First, Japan is faced with the situation in which North Korean shorter-range missiles—immune to US homeland-based BMD—could threaten or attack Japan independent of conflict on the peninsula. The national missile defense in the United States is capable only for protecting the United States from attacks. Second, defenses make sense considering Japanese domestic politics (e.g., constitutional interpretations placing constraints on offensive military capabilities). Third, BMD also enables autonomous protection of Japan's people and its interests, particularly against ballistic missile-based coercion like that from North Korea.

Deterrence has been a central concept within Japan's national security and military strategy for years. Recently, it was restated in various government security and budgetary documents. For example, in *Defense of Japan 2017*, more than thirty pages deal explicitly with deterrence. In that section, technologies, capabilities, and procedures and tactics are all expressions of deterrence by denial, including air and naval warning and surveillance such as anti-submarine and air systems; active defense technologies such as BMD, naval coastal defense patrols, and air defense scrambles; means of rapid mobilization; consequence management capabilities and practices; civil defense procedures; means of redundancy across command and control, cyberspace, and space assets; and, technologically partnering with the United States on the Aegis-equipped SM-3 Block IIA advanced BMD interceptor. The 2018 version of this document reflects deterrence—including BMD—as Japan's central military strategy whereas regional stability is its key political strategy. And, in Japan's 2019 defense budget, mobilization capabilities—another form of denial— were called out as deterrence enhancements.

Japan's indigenously operated BMD now represents the second best in the world, behind only the US system (though Dima Adamsky's claim in this volume of Israeli prowess in this area should likewise be considered).[23] Japan's layered BMD system includes both upper-tier defense and lower-tier defense technologies capable of engaging ballistic missiles at various distances and altitudes. Upper-tier defense, able to intercept warheads while still in space, is currently provided by six destroyers equipped with the Aegis launch system, with two more under construction. All will be upgraded with the extended range SM-3 Block IIA interceptor. Lower-tier point defense is provided by the Patriot system with PAC-3 missiles. Patriot launch batteries are deployed across Japan. In the past couple years, Japan has also considered purchasing the Terminal High Altitude Area Defense (THAAD) system to provide cutting-edge area defense technology at ranges and altitudes between Patriot and Aegis, which would give Japan a third layer.[24] In the future, Japan will also acquire the land-based Aegis Ashore system. Tokyo is the primary area of defense.[25] Japan's BMD system is also "interoperable" with the US BMD system through radar detection capabilities based in Japan.[26] Having begun deployment in 2007, Japan's BMD is expected to grow to eight Aegis BMD ships and seven PAC-3 firing units.[27] This could mean more than 700 SM-3 missiles on Aegis and more than 500 PAC-3 missiles on Patriot launchers, providing Japan a capability to simultaneously defend against potentially dozens of attacking ballistic missiles.[28] Tactically, the flight time of a North Korean offensive ballistic missile to Japan is about 10 minutes, compressing time available for decision-making and defense.[29] For this reason, Japanese leaders have provided delegation of authority to launch BMD assets below the political leadership to military commanders.

JAPAN'S BMD DETERRENCE STRATEGY VS. DPRK

While there are various ways to approach or analyze the deterrent dynamics between Japan's BMD and North Korea, it is organized into three categories: the strategic level, the operational level, and the psychological level. One will find overlap among these categories though this should be expected. Japan's active defense BMD technology, while not deterring every single North Korean missile test over Japan, has in the whole demonstrated significant deterrence-by-denial value against coercion below the threshold of armed conflict.

Denial at the Strategic Level

For DPRK in the post–Cold War era, national survival relies on ballistic missiles and deterrence, whereas coercion using those missiles is applied to incrementally affect North Korea's relative position and well-being. Likewise, Japan's deterrence strategy is essentially a counter-coercion deterrence strategy particularly regarding threats below the conflict threshold. In theory and practice, Japan's recent and ongoing deterrence strategy vis-à-vis DPRK coercion is chiefly one of denial. At the strategic level, BMD deters by denying the North the full efficacy of its underlying, if not blunt, instrument of coercion.[30] Japan postured BMD technology opposite DPRK ballistic missiles. Japanese BMD and demonstrations of resolve with its BMD deny North Korea its perceived political and economic benefits by reducing the coercive power of North Korea's ballistic missiles over Japanese policy. Undermining the value of DPRK's ballistic missile force—the military instrument with sufficient range to threaten Japan—is essential in countering DPRK coercion. North Korea simply has not developed any other military, political, or economic lever through which it can coerce Japan on the scale or destructive power of ballistic missiles. This is made possible with an effective layered BMD system. Rendering ballistic missiles far less effective denies their perceived benefits and, therefore, undermines North Korea's power of coercion. As a result, Japan is not only safer through the defense offered by active defense technology but is in a stronger political position.

This denial characteristic of Japan's BMD can also be considered in terms of DPRK leadership strategic risk propensity. North Korean behavior regarding Japan's active defense technology development and deployment generally reflected a move from greater risk tolerance in the Cold War period of the North's strategy of confrontation to one of relative risk aversion in the period of ballistic missile-dominated coercion, the latter of which paralleled the development of Japan's BMD in the 1990s and its maturing including operational deployment in 2007 and being postured to engage provocative North Korean missile tests in 2009 and 2017. Despite the challenging nature of Japan's BMD program—and quite possibly because of it—the risk tolerance of DPRK leaders toward Japan in DPRK's overall coercive strategy was lower than that toward ROK since a more violent course of action against ROK was apparent.[31]

Denial at the Operational Level

DPRK's ballistic missile force is a vast capability not simply of utility for warfighting purposes during conflict. Likewise, Japan's BMD has deterrent value against threatening ballistic missiles under conditions short of war. For example, BMD defends against feared DPRK "cheap shot" missile raid attacks that could occur well below the threshold of war and US conventional or nuclear retaliation provided under extended deterrence security guarantees. Japan's BMD deters these types of raids because BMD denies the operational effectiveness DPRK could reasonably hope to achieve with its ballistic missiles in such raids—such as damage on Japanese leadership, military, industrial, or financial targets—or a reasonable level of post-raid escalation with ballistic missiles. This is the advantage of having a technologically superior active defense system.

Multilayered BMD can also have denial effects on DPRK operational ballistic missile tests. For example, during Kim Jong-il's reign, Japan's BMD program during the period immediately following North Korea's 1998 TD-1 launch (September 1998–November 2003) led to a period of political "warming" when DPRK's leaders may have sought to stifle support among Japan's public and legislature for large investments in BMD, which would have ultimately helped North Korea achieve better outcomes from Japan. Further, the Sohae Satellite Launching Station began construction in 2001 during this warming period as North Korea was clearly looking for a way to test missiles in another direction away from Japan.[32] Additionally, on April 5, 2009, a longer-range variant of Taepodong was launched, but the North Koreans opted to comply with a United Nations request for safety of flight and navigation prelaunch notifications. In the lead up to the launch Japan gave orders to field its BMD assets, defend Tokyo and other areas, and prepare to shoot the missile down.[33] The behavior and statements of North Korea for the 2009 event were clearly different from the 1998 surprise launch—indicating North Korea changed the missile's flight profile to one that was less threatening to Japan and in response to awareness of Japan's operational BMD deployment. If so, Japan achieved a deterrent effect for which its BMD was designed. It did not deter the launch; that was not likely the intent. Rather, DPRK political and coercive benefits hoped for in testing a ballistic missile in a threatening or provocative manner were denied by influencing North Korean leadership to test the missile in a different operational configuration: North Korean leaders altered their political and military behavior to one less provocative to Japanese leaders and less dangerous to Japan's population.

Such denial advantages of Japan's BMD have continued during the Kim Jong-un era. Since Kim Jong-un took control of North Korea, the DPRK carried out over 80 ballistic missile tests, a noticeable increase in frequency from that of the Kim Jong-il era likely as Kim Jong-un adopted an approach more tolerant of missile failures. Many of these missiles flew lofted trajectories with short ground tracks or otherwise flew short of or away from Japan when they could have flown over Japan. However, only two missile tests (both in late 2017) ground tracked over Japanese territory similar to the 1998 and 2009 tests. These two tests represent only 2.4% of these high-frequency tests (two of eighty-four tests). By comparison, in the missile testing era of his father, Kim Jong-il, 4.3% (two of forty-seven tests) of DPRK missile tests ground tracked over Japanese territory, itself a very small fragment.[34] As a percentage, this is a reduction of coercive tests over Japan and suggests considerable North Korean restraint, particularly in the recent high-volume testing period. That is consistent with restrictive deterrence principles because they represent not only a low frequency of such tests but also because flight profiles were constrained to very high altitude when ground tracking over Japan. And this despite having more missile types and actual missile tests that can traverse Japanese territory than in the earlier Kim Jong-il era. When further analyzing the two 2017 tests, it is shown that they flew at an apogee above the threshold of legal airspace, likely a North Korean leadership calculation owing to Japan's BMD capabilities.[35] Further, these were of the Hwangsong-12 missile designed with range to strike US targets on its territory of Guam to the south, not targets in Japan.[36] As such, they did not appear to reflect direct acts of coercion against Japan. If, for example, North Korea's leaders had calculated that Japan would acquiesce from its defiant position or there would be no significant lasting consequences from Japan for these tests they calculated in error: while Japan chose not to engage either of the two missiles, its position against North Korea not only stiffened but immediately following the second test Japan decided to purchase from the United States, at a cost of over $2 billion, two "Aegis Ashore" systems with advanced SM3 Block IIA interceptors to complement its shipborne Aegis systems.[37] While not adding another layer to Japan's BMD system, this new technology will greatly enhance Japan's ability to engage a larger number of enemy warheads at great distance further reducing the coercive and operational efficacy of North Korea's ballistic missile force against Japan. The only variable capable of significantly altering North Korean decision-making regarding ballistic missile flights over Japan was Japan's high-technology active defense BMD system.

Denial at the Psychological Level

DPRK national and leadership psychological factors are also important deterrence considerations in the Japan-DPRK case. The DPRK's political, cultural, and identity factors include such things as securing state sovereignty and protection from intervention and occupation, national honor, self-reliance, technological advancement, and its historical place in regional and world history. For example, it is possible that Japan's BMD helped convince North Korea's leaders to sanction a ballistic missile test moratorium for a time, possibly believing operational and political benefits from the tests would be sacrificed should their missiles be shot down by Japan. Some suggest North Korea's leaders sought to maximize the probability of coercion's success by *appearing* dangerous if not unpredictable. For example, Derek Smith argued that North Korea manipulated

US and allied fears of North Korea's "rogue" state leadership irrationality and impacted US behavior.[38] BMD can deny the psychological utility of this tactic precisely because of the wholly defensive nature of Japan's BMD and overall defensive military posture vis-à-vis the DPRK and its ballistic missiles. Further, even though North Korea changed behavioral patterns in its 2009 TD-2 missile test and arced its 2017 tests high over Japanese territory, if Japan would have opted to shoot down the missile, it may have changed North Korea's leaders' image at home and abroad sufficiently to lead them to violent provocation with Japan. Japan's choice to employ BMD but withhold from shooting at missiles allowed DPRK leaders to "save face." They may have demonstrated strength to the world with a very small percentage of missile tests over Japan, though they compromised in how they carried out those tests. Having the missile shot down may have been the worst outcome for North Korean leaders: being outdone technically (and militarily) by Japan; tarnishing their domestic image with all audiences; and a direct external challenge to their source of sovereign and coercive strength (i.e., ballistic missiles with escalatory threats of WMD), which would demand a response and possible escalation.[39]

Conclusion

Japan's development of sophisticated BMD active defense technologies intended, in significant part, to deter DPRK coercion should be seen in context of negative deterrence consequences of Japan's BMD choices along with proliferation concerns both from DPRK and Japan. First, Japan's successful use of BMD deployment might push North Korea to employ other means of coercion for which Japan is not yet fully prepared to defend itself such as a robust DPRK offensive cyber capability that could wreak havoc upon Japan's critical infrastructure, financial networks, dependencies in space, or its military forces and command and control. Second, North Korea could choose to develop BMD countermeasures or build more ballistic missiles to overwhelm Japan's BMD. While successful overall in deterring DPRK behavior and coercion, Japan's BMD deter-rence-by-denial approach did not stop DPRK's ballistic missile production and deployment—DPRK choices taken most likely as countermeasures to Japan's expanding BMD. DPRK leaders likely reasoned they simply needed more ballistic missiles to overwhelm Japan's BMD for coercion to work, and in war if needed. And, given Japan's decision to expand its BMD system with Aegis Ashore batteries that will further confound North Korean ballistic missile targeting schemes, North Korean leaders may choose to expand its ballistic missile deployments even further. Third, the prospects of Japan developing offensive capabilities is possible. For example, Japan's Defense Ministry formally addressed the option of acquiring strike capabilities to attack DPRK ballistic missile bases. Part of the rationale given by Japan's Minister of Defense, Itsunori Onodera, was that Japan's BMD system could allow some missiles attacking Japan to "leak through" creating dire consequences in Japan, suggesting Japan's confidence in active defense technologies to deter might not be foolproof.[40]

In general, however, Japan's BMD technology has proven to be an effective counter-coercion deterrence-by-denial capability below the threshold of armed conflict, given that DPRK coercion has been based on ballistic missiles and the threat of WMD escalation. In fact, the political stability of the Japan-DPRK relationship is strengthened by, if not founded on, Japan's BMD technology and the deterrence-by-denial influence it provides. Japan's BMD operated as designed in a restrictive deterrence perspective. For example, DPRK conducted many missile flight tests over the past three decades, but the number of those that overflew Japan proper was minimal and the frequency of those overflying Japan actually went down in the Kim Jong-un era compared to the former Kim Il-sung era. Further, latter tests that did overfly Japan were less provocative, having flown farther north, at higher altitudes, and with advance international notices. In addition, while there were many more tests in recent years, many flew short, southerly, or lofted trajectories avoiding overflight of Japan altogether. The DPRK still has ballistic missiles, but they are less effective now in coercing Japan. Japan's BMD also contributed to reducing DPRK conflictual or provocative behavior toward Japan and reassured Japan's population and prepared for defense of the nation in wartime should that occur. Moreover, other states across the globe are also buying into the denial value of BMD and are buying or developing BMD technologies.[41]

Japan's BMD capability has also performed consistent with theoretic notions of deterrence by denial through use of active defense technology. While defensive in nature, Japan's BMD denies DPRK efficacy of its coercive strategy, thus deterring coercion in practice. The Japan-DPRK case, however, is a unique deterrence-by-denial case. This is true not only because of Japan's use of exclusively defensive capabilities to provide such deterrence but also because the object of deterring influence is DPRK coercive behavior in the sub-conflict phases associated with general deterrence. This has made deterrence-by-denial operative through active defense technology below the threshold of conflict, rather than simply relying upon legacy deterrence through threats of cost imposition to avoid the breakout of war.

Notes

1. Throughout this chapter, the acronym DPRK is used interchangeably with the common reference North Korea to identify the same state actor.
2. Defense Intelligence Agency, "Ballistic and Cruise Missile Threat 2017"; Schiller, *Characterizing the North Korean Nuclear Missile Threat*, 62.
3. Reportedly, the North Koreans asked for $10–20 billion from Japan. Manyin, *North Korea-Japan Relations*, 3.
4. This is expressly stated in the DoD's *Missile Defense Review*, 27–29.
5. Huth and Russett, "General Deterrence between Enduring Rivals," 61–73. General deterrence was simply conceived as steady state conditions between two actors short of crisis or conflict.
6. Reference to Japan's BMD speaks to Japan's multilayered homeland defense system.
7. Another Japanese active defense capability, the F-35 stealth fighter is also one of the highest technology capabilities in the world that can serve deterrence by denial purposes.
8. A longer U.S. Department of Defense definition suggests deterrence operations are "integrated, systematic efforts to exercise decisive influence over adversaries' decision-making calculus in peacetime, crisis, and war to achieve deterrence." Department of Defense, *Quadrennial Roles and Missions Review Report*, 5.
9. In addition to cost imposition and denial, an adversary's decision calculus is informed by the perceived costs and benefits of restraint. That is, the deterrer seeks to mitigate the adversary's perceived costs of restraint while incentivizing his benefits of restraint (also dubbed the "carrot" of deterrence).
10. Snyder, *Deterrence and Defense*, 4, 9–11, and 14–15.
11. On the idea of "deterrent defense" see Schelling, *Arms and Influence* (2nd ed.), 78–79.
12. On the broad distinctions between restrictive, absolute, and specific deterrence, see, for example, Howell, "The Restrictive Deterrent Effect"; Jacques and Allen, "Bentham's Sanction Typology." Restrictive deterrence could also apply to cyber defense technologies which cannot reasonably be expected to deny all adversary cyberattacks but could be expected to modify adversary behavior including the frequency, intensity, and scope of attacks by denying those deciding to carry out such

attacks the full range of benefits sought and by complicating cyberattack planning.

13. See both the *Missile Defense Review* (2019), 29, and its predecessor Department of Defense, *Ballistic Missile Defense Review Report* (2010), 11.

14. *Fact Sheet: Missile Defense and Deterrence* (Washington: U.S. Department of State, 2001). See the heading entitled, "Emerging Threats and the Need to Diversify our Approach to Deterrence."

15. Past DPRK behavior has included violent acts such as the sinking of the South Korean naval vessel *Cheonon*, artillery strikes into South Korea, naval incidents with Japan and South Korea, and missile and nuclear tests.

16. Pinkston, *The North Korean Ballistic Missile Program*, 47.

17. Pinkston, *The North Korean Ballistic Missile Program*, 47.

18. Bluth, *Korea*, 161–162.

19. Takahashi, "Ballistic Missile Defense in Japan," 23.

20. According to Cha, "The United States has failed for over twenty years to deter DPRK development and testing of its ballistic missiles." Cha, *The Impossible State*, 223–224.

21. Perry, Davis, Schoff, and Yoshihara, *Alliance Diversification*, 144.

22. Park, "Japanese Strategic Thinking toward Korea," 192.

23. Dawson, "Japan Shows Off Its Missile-Defense System." The number two ranking was based on system sophistication or quality. See Chapter 7 of this volume.

24. Mizokami, "Everything You Need to Know."

25. Kang and Lee, "Japan-Korea Relations."

26. *Fact Sheet: Missile Defense and Deterrence*, 9. According to Abmann, mobile sea-based platforms, such as Aegis, are optimum for Japan given its island chain geography. Abmann, *Theater Missile Defense*, 133–137.

27. Japanese Ministry of Defense, *Defense of Japan 2017*, 318–350; Missile Defense Advocacy Alliance, "Making the World a Safer Place: Japan"; Associated Press, "Japan deploys PAC-3 missile interceptor"; Japanese Ministry of Defense, *National Defense Program Guidelines for FY 2014 and beyond*, 31.

28. The figures here have been assimilated from multiple sources reflecting maximum potential and may not reflect the realities of Japan's actual BMD interceptor force or the capacity of that force to engage North Korean attacking ballistic missiles.

29. Kaneda, Tajima, Kobayashi, and Tosaki, *Japan's Missile Defense*, 92.

30. Japan's BMD system is explicitly identified by the Government of Japan as part of its "deterrent" capability. For an example see *Defense of Japan 2017; National Defense Program Guidelines for FY 2011 and beyond,* 10.
31. For example, DPRK's torpedo attack on the ROK *Cheonon* on March 26, 2010, that killed forty-six ROK sailors, and its artillery shelling on November 23, 2010, of Yeonpyeong Island that killed four and wounded dozens.
32. Nuclear Threat Initiative, *Sohae Satellite Launching Station.*
33. Japan Ministry of Defense, *Order for Operation of the Self-Defense Forces.* This order was issued in response to advance notice provided on March 12, 2009, by North Korea to the UN.
34. On DPRK missile test numbers, see Center for Strategic and International Studies, "North Korean Missile Launches & Nuclear Tests"; CBS News, "North Korea missile tests"; McCarthy, "The Worrying Escalation." There were no DPRK ballistic missile tests in 2018.
35. These trajectories were beyond Japan's PAC-3 coverage but within its Aegis engagement capability.
36. Hwasong-12 tests on August 29 and September 15, 2017. See *The Straits Times,* "North Korea's Hwasong-12 is the 'most serious missile to watch'"; Analysans, "North Korea's September 15 Hwasong-12 Test; Elleman, "North Korea's Hwasong-12 Launch."
37. Yamaguchi, "Japan to buy Aegis Ashore."
38. Smith, *Deterring America,* 68, 71–74.
39. Other psychological perceptions, however, might suggest perceived DPRK benefits were not denied. For example, Japan's BMD did not deny the North Korean people their sense of national pride in DPRK's ballistic missile programs or occasionally their emotive satisfaction in relations with "imperial" Japan. BMD did not deny DPRK the means or benefits of nuclear testing. Neither did BMD deny DPRK the option to simply build more ballistic missiles or create alternative means of coercion such as offensive cyber capabilities.
40. See, for example, Reuters, "As North Korea missile threat grows."
41. In addition to ROK other notable cases include Israel, GCC states, and Taiwan. India, too, is developing a layered BMD system to strengthen deterrence by defending its leadership and nuclear command and control capabilities, in this case providing it a second-strike capability. See O'Donnell and Joshi, "India's Missile Defense"; see the section, "Why India Wants a BMD System."

CHAPTER 7

DETERRENCE BY DENIAL IN ISRAELI STRATEGIC THINKING

Dmitry (Dima) Adamsky

This chapter explores the shifts in Israeli thinking about deterrence and examines the role that deterrence by denial currently plays in Israeli security policy. In Israel, deterrence has been, and still is, one of the pillars of its national security concept, together with early warning, battlefield decision, and, recently, defense. Israelis have been focused on the concept of deterrence and continuously practiced deterrence strategy in their national security. The concept was incepted and articulated in the 1940s and 1950s by the founding fathers of the state. Because the traditional Israeli military doctrine was driven by the "cult of the offensive," deterrence was seen either as a by-product of the battlefield decision against the state actors, or as a result of retaliation against non-state enemies. In both cases, it was primarily *deterrence by punishment*, and not deterrence by denial that, together with a defensive form of warfare, was discredited as irrelevant and traditionally disregarded.

In the last two decades, in framing the transformation of the Israeli national security concept, defense and deterrence by denial has gained more traction. The shift was driven partially by the greater orientation to defense in the IDF concept of operations, weapons systems procurement, and organizational structures, as lessons were learned from the Yom Kippur War and the First Gulf War. In part it was internalizing the change in the nature of the main threats that migrated from conventional to nonconventional and sub-conventional battlefields. The lessons learned during the last decades from the campaigns against asymmetrical enemies employing terrorism and guerrilla tactics also contributed to this transformation. The growing attention paid to deterrence by denial, which deviates from traditional practices in Israeli national security, is the central focus of this chapter. Despite the still-dominant role of offense and deterrence by punishment in Israeli strategic thought, deterrence by denial has taken root and shows promise for managing contemporary Israeli national security threats.

The first section of this chapter outlines the traditional Israeli deterrence paradigm. The second section examines the changing nature of threats, the Israeli understanding of them, and the subsequent paradigmatic change that the Israeli approach to deterrence underwent in the last decade. The third section discusses the growing importance attributed to deterrence by denial that occurred as part of this change and examines the balance between it and the traditionally dominant deterrence-by-punishment strategy in the current Israeli national security paradigm. The conclusion reflects on the factors that hamper further integration of deterrence by denial into the Israeli security practice.

TRADITIONAL ISRAELI DETERRENCE PARADIGM

The Israeli concept and practice of deterrence are unique; they differ from the canonical Western deterrence theories, as Patrick Morgan, Alex Wilner, and James Wirtz, among others, have highlighted in this volume.[1] The main difference is on the question of the use of force. Classical deterrence theory postulates that resolve to employ force demonstrates a shift from a policy of influence to a policy of control. Consequently, it is a symptom of deterrence failure. In contrast, in the case of Israel, use of force is not a symptom of deterrence failure but rather an integral component of this strategy. Several reasons account for these varying conceptualizations of deterrence.

The first reason is the distinct empirical-theoretical contexts in which thinking about deterrence evolved in both countries (i.e., the United States and Israel). In the United States, classical deterrence theory evolved against the backdrop of the Cold War's nuclear standoff. The Israeli conceptualization and practice of deterrence, in contrast, emerged in a different empirical context—actual military interactions with state and non-state actors.[2] Different empirical contexts demanded different theoretical approaches. American nuclear strategists conceptualized deterrence as use of a threat aimed to influence an opponent's strategic calculus.[3] Whether the threat is used to preserve *status quo* (deterrence) or to change it (compellence), it is a strategy of coercion, not of brute force. An opponent's strategic behavior is shaped by psychological influence, not by physical control. Thus, if according to classical theory, the aim of deterrence is the prevention of an action through fear of consequences, then use of force either by the subject or by the object of deterrence is a symptom of deterrence failure.

The Israeli approach has been in dissonance with the strategic bombing school of thought, but it resonates with criminological conceptualizations of deterrence, as expanded by Janice Gross Stein and Ron Levi.[4] Originating in the conventional empirical context, in which deterrence by definition is contestable, the Israeli approach rejects the notion of "absolute deterrence." The use of force is tolerable in non-nuclear interactions, and Israeli deterrence has been continuously tested.[5] Consequently, use of force is not a failure but an integral component of deterrence, necessary for the success of strategic education effects.[6] This fits with the Israeli national security concept, which expects continuous rounds of violence and sees the next military clash happening only in a matter of time. Israeli deterrence thus does not aim to eliminate violence but rather to postpone it and reduce its magnitude (Jonathan Trexel explores this in his chapter on Japanese deterrence posture vis-à-vis North Korea).[7] As in criminology, where punishment is used to construct and preserve rules of conduct, the Israelis saw in military force an instrument to establish and maintain norms of strategic behavior.[8] Deterrence thus becomes a series of forceful acts aimed at educating challengers about the "rules of the game" and to force them and other actors "to internalize these lessons" over an extended period of time.[9] For this reason, the Israeli approach relates more to deterrence as conceptualized by law theory.

The second reason relates to different attitudes regarding capability and resolve in both cases. In the nuclear realm, deterrence failure is Armageddon. In the conventional realm, it is a "learning-teaching" episode about capability and resolve.[10] Consequently, mainstream Western (nuclear) deterrence theory was primarily preoccupied with the problem of stability—how to avoid turning credible threats into a provocative crisis that undermines the *status quo* and results in war. In contrast, (conventional) deterrence saw stability as a lesser concern than credibility due to the acceptable costs of warfare.[11] In nuclear deterrence, credibility of capability was unquestionable; it was the resolve to use nuclear weapons that was questioned. In conventional deterrence, in contrast, the credibility question focused on capability. Can a military machine deny aggression, punish effectively, and ensure victory in military terms?[12] Israeli strategists traditionally assumed that episodic use of force would demonstrate capability, maintain a reputation for toughness, and assure credibility.[13]

As a result of these peculiarities, Israeli deterrence categories differ from the canonical ones, including those discussed by the other contributors to this volume.[14] Although the Israeli conceptualization of deterrence is not codified doctrinally, it skillfully reflects strategic complexity. Janice Gross Stein was the first to establish this point. She and Elli Lieberman demonstrate that deterrence is seen in Israel not in a dichotomist manner but a nuanced one—while one type of deterrence collapses, others may stay intact. General deterrence may be preserved, but specific and current types may fail. Stability may be established despite motivation to challenge deterrence.[15] After synthesizing available Israeli sources, Uri Bar-Joseph categorizes Israeli deterrence by goals: *current deterrence*, which prevents violence and escalation of the low-intensity conflict; *specific deterrence*, which prevents limited moves that endanger vital interests; *strategic deterrence*, which prevents a general war;[16] and *cumulative deterrence*, which aims to persuade enemies that attempts to achieve their goals on the battlefield are doomed. [17] Boaz Atzili and Wendy Pearlman have added a category to this typology by means. In addition to deterrence by denial and deterrence by punishment, Israel practices *triadic deterrence*— influencing non-state proxies through their state patrons.[18] Known as a "leverages approach," this and the aforementioned forms of deterrence have been practiced since Israel's establishment.

The traditional Israeli deterrence mechanism is best understood as part of the Israeli unwritten national security concept. Known as the "security triangle," this collection of stratagems introduced by its founding fathers assumes that Israel would always engage its enemies from an inferior position in terms of territory, human, and natural resources, and tolerance to casualties and to international pressure. This imbalance would prevent Israel from achieving political goals and terminal strategic victories by military means; the operational, battlefield decision would be the maximal achievement. In an indefinite number of clashes, the enemies could afford to lose in every round, but a major operational defeat would mean Israel's annihilation. Thus, operationally, Israel should be victorious in every round of warfare. As a result, a *status quo* country with defensive strategy, Israel adopted offensive battlefield tactics and rested its national security concept on three pillars: deterrence, early warning, and battlefield decision. Deterrence preserves stability and extends the duration of peace, but if it fails, then the second pillar—early warning about impending aggression comes into play. It provides the necessary time to deploy the pool of reservists while a small standing force either strikes preventively or blocks enemies' initial advance. Then, the third pillar—battlefield decision—materializes. Although not a terminal strategic knockout, it should produce decisive operational victory at this round of conflict. According to the traditional Israeli culture of war, only an aggressive, maneuverable warfare swiftly penetrating the adversary's rear may produce this result.

In addition to solving the burning operational problem, battlefield decision aims to recharge the batteries of deterrence and postpone the next round of violence. Thus, deterrence is a by-product of battlefield decision.[19] Traditional Israeli strategic thought distinguished between two main challenges—fundamental security threats (*bitakhon bsisi*) and current security challenges of lower magnitude (*batash*). Deterrence of fundamental threats has been an automatic by-product of battlefield decision and was achieved by "default." In contrast, deterrence of current threats was maintained through retaliation operations that had been "designed" for this purpose.[20] Israelis did not expect deterrence to work forever; they presumed an indefinite number of its failures and a constant need to maintain it.[21]

This unique national security concept produced a peculiar Israeli deterrence *modus operandi*. First, in keeping with the "cult of the offensive," Israel has been inclined to practice deterrence by punishment, rather than deterrence by denial.[22] Second, driven by a quest for absolute security,[23] the Israeli approach is essentially asymmetrical. Traditionally, it was assumed that a deterrence regime can be achieved only from a position of superiority, not of parity; Israel sought to deter without being deterred.[24] Third, the Israeli strategic community has been operating under the assumption that declaratory threats are insufficient. The sword by itself does not establish credibility; it should be constantly bloodied to maintain deterrence. Instrumental signals—periodical execution of threats—were seen as essential to communicate resolve and capability. In order to maintain a reputation for toughness to prevent future attacks, Israel periodically resolves to employ force to send costly signals.[25] This is a "serial deterrence"[26] of "unlimited use of limited force."[27] Finally, striving for escalation dominance and excessive and disproportional use of force "beyond expectations of the enemy" characterizes the Israeli approach to deterrence and compellence,[28] two terms that Israeli security jargon frequently conflates.[29]

In sum, in contrast to classical Western deterrence theory, the traditional Israeli approach *a priori* expected deterrence failure, was less concerned with the question of stability, was preoccupied with the issue of credibility, mostly focused on the capability component, and periodically used limited force to maintain its deterrence regime, thus leaning on punishment rather than on denial.

LEARNING ABOUT CHANGE IN THE NATURE OF THREAT AND PARADIGMATIC SHIFT

This traditional deterrence paradigm was intact up to the mid 1990s—until the Israeli strategic community internalized its flaws and identified the change in the nature of the threat. These understandings evolved into a crisis of knowledge and eventually resulted in a shift in the Israeli deterrence paradigm.

First, a growing belief emerged that there was no causal link between battlefield decision and deterrence effect. According to the traditional Israeli wisdom, ultimate victory reinvigorates deterrence whereas poor battlefield performance bankrupts deterrence and emboldens enemies.[30] However, accumulated historical evidence was running against this assumption. Enemies initiated wars when Israel was at the heights of its military power—in 1967, 1968, 1973, and in both intifadas—and avoided such moves when Israel's military reputation had shrunk. Israeli strategists gradually realized that aggressive motivations were affected less by the correlation of forces, and more by the quality of the *status quo* and the balance of interests.[31] This insight undermined the logic behind the traditional deterrence mechanism.

Second, Israeli strategists identified the change in the character of war and in the nature of the threat. Since the late 1990s, they started to question their ability to make battlefield decisions. Traditionally, ultimate victories demonstrated indisputable Israeli capability and resolve. The best way for the IDF to demonstrate its unquestionable *blitzkrieg* superiority was to defeat conventional armies in a massive operational engagement. However, this *modus operandi* produced an unforeseen outcome. Indisputable IDF superiority deterred its enemies from competing on the conventional battleground and forced them to wage war where Israel's military advantage was nullified—sub-conventional and unconventional arenas. On these asymmetrical battlegrounds, the IDF was losing its ability to produce clear-cut operational victories.

This "designing around" phenomenon[32] materialized in two asymmetrical "theories of victory"[33]—terror and WMD—which are underlined by a common denominator—ballistic capabilities. During the 1990s, the dominant tactics that Israel faced from its enemies were insurgency, suicide terrorism, development of WMD, and acquisition of high-trajectory arsenals. The "ballistic *muqawama*" (resistance) trend gathered momentum during the 1990s and is, as of this writing, the kernel of state and non-state actors' theory of victory. Regional conventional armies and hybrid military forces structured themselves not to invade Israel but to defend their ballistic capabilities. To offset Israeli military superiority, opponents based their theory of victory on "absorption," aimed at downgrading the utility of Israel's precision guided munitions; "deterring" the IDF from fighting, by shifting warfare to problematic dimensions; and on winning war by not losing, thus creating "attrition."[34] By the early 2000s, the IDF found itself on an unanticipated battlefield and started to question its ability to materialize classical battlefield decision. Oriented to destroy conventional armies, the Israeli war machine was less relevant for these new challenges. It took Israeli practitioners about a decade to internalize that asymmetry had become the strategy of their principal adversaries.

The first insight made Israeli practitioners question whether deterrence stems from battlefield decision. The second insight suggested that the changing character of war had incapacitated the IDF from generating one. Underscored by Israel's quest for international legitimacy, especially against the backdrop of criticism following Operation Cast Lead,[35] these insights further limited its resolve to employ force. Internalization of these two problems gathered momentum during the 1900s and stimulated critical thinking about battlefield decision and deterrence concepts. Somewhere following the First Intifada, the IDF started to experience a crisis of knowledge when senior commanders began to intuitively realize that battlefield decision does not automatically produce deterrence. Trial and error and lessons from Operation Accountability (1993) and Operation Grapes of Wrath (1996) resulted in a reorientation of military operations from deciding towards influencing the opponent. The mainstream assumption of IDF planners became that the enemy should be contained and managed but could not be defeated. [36]

During the Second Intifada, against the backdrop of acknowledging the difficulty of defeating the opponent in decisive operations, the IDF General Staff started to seek "substitutes to battlefield decision"—military operations that would produce tangible results under the IDF's objective limitations. "Decisive operations," did not disappear, as exemplified by Operation Defensive Wall (2002), but became an exception to the rule.[37] Staff work on "substitutes to decision," was a prolongation of "influence operations" which had been incepted during the 1990s in Lebanon and further stimulated by combat experience during the 2000s. The IDF sought a new paradigm especially as lessons from the 2006 war demonstrated that declaratory threats prior to the war were not effective and that employment of force in reaction to aggression is not really deterrence.[38] Lessons learned from the Second Intifada, the 2006 War, and Operation Cast Lead (2009) resulted in a more systematic thinking about the "substitutes to decision," and matured into two concepts: deterrence operations and *mabam* (campaign between the wars). These two interrelated concepts, in a nutshell, represent a new Israeli deterrence paradigm.

Under this new paradigm, deterrence is not an outcome of an ultimate victory in a major war but a cumulative result of intensive short operations and ongoing limited engagements in between major wars. This resonates with the traditional Israeli *batash* deterrence approach and may sound like old wine in new bottles. However, Israeli practitioners argue that this is a new paradigm. To them, the traditional *batash* deterrence model has become the central mode of operations against state and non-state actors. They assert that Israel seeks to produce deterrence by "design," through initiated operations, and not by "default," through the outcome of a major war, as it used to do. According to them, "deterrence operation" and not "decisive campaign" has become the main course of action to ensure security, stability, and deterrence. Under this new paradigm, when Israel estimates that deterrence vis-à-vis a particular actor is evaporating, it initiates operations like Operation Cast Lead and Operation Pillar of Defense.[39]

Against any opponent, deterrence is not an outcome of decisive victory in a major war, not a stand-alone surgical strike, but an uninterrupted and continuous process. This campaign has a unified operational logic that knits together diffused employments of force of varying magnitude over time and space. A deterrence operation is a last resort aimed to restore an evaporating deterrence regime vis-à-vis a particular actor. This thinking again strongly resonated with the traditional "criminological" approach to deterrence, as Stein and Levi highlight in chapter 3, that assumes that violations in a particular jurisdiction are unlikely to go to zero but argues that punitive action precedents lead offenders to reevaluate their intentions, thus reducing the rate of misbehavior.[40]

To avoid this less-desired, higher-intensity action, the IDF practices low-intensity *mabam* to maintain a deterrence regime and prevent its evaporation in between deterrence operations. In principle, *mabam* aims to create trends and conditions under which the geostrategic situation does not demand a deterrence operation, *à la* Operation Cast Lead or Operation Pillar of Defense. *Mabam* evolved out of special operations logic during the 2000s and is another variation on the "deterrence" theme.[41] It became the IDF concept of operations for the period in between acts of intensive warfare—the most significant time span of the IDF for which it previously did not have an operational mode of conduct. *Mabam* is a joint project of the Israeli strategic community and consists of an ongoing series of special operations against an opponent's valuable military assets and of forceful preventions from its acquisition of advanced military capabilities. These clandestine attacks take place during the peacetime routine—in between major clashes (i.e., "deterrence operations"). *Mabam* aims to force the adversary to postpone the next round of warfare and to reduce the magnitude of future attack. As opposed to "decisive battle," *mabam* does not have an end state, and its goal is to constantly preserve the balance of regional trends in Israel's favor, what eventually results in deterrence. *Mabam* seeks to leave a psychological imprint on the adversary, to increase his feeling of vulnerability, to dissuade him from undesired actions, and to enable better opening conditions in the next round of warfare. Strikes on various targets across the Middle East that have been attributed to Israel during recent years presumably constitute examples of *mabam* in action.[42]

Deterrence operations and *mabam* reflect both continuity and change in Israeli strategic thought. They reflect continuity since they continue to see offensive employment of force (punishment) as an integral component of deterrence and further refine the deterrence-by-design approach practiced in *batash* missions. They reflect change in a sense that they moved away from deterrence by default that had been practiced in conventional warfare and adopted deterrence by design as their main *modus operandi*. In addition, the change has been evident in the greater emphasis in the Israeli approach on defensive mode of operations that subsequently increased the role of deterrence by denial.

DETERRENCE BY DENIAL IN ISRAELI STRATEGY

The shift in the deterrence paradigm coincided with an additional trend in the Israeli strategic thinking and policy. Several defensive capabilities and defense-related activities that strategic studies traditionally associate with deterrence by denial started to gather momentum. Today, deterrence by denial plays a role bigger than it ever has in Israeli national security policy. How did deterrence by denial acquire such unprecedented importance and take over several missions that were previously in the jurisdiction of deterrence by punishment? The following interrelated processes account for this shift.

The first impetus to move away from the overall emphasize on offensive mobility, and toward an increasing attention to defense, evolved in the Israeli strategic thought following the lessons learned from the Yom Kippur War and the First Gulf War. In as much as static defense totally discredited itself in 1973, the notion of opting for reactive defense based on advanced IT technologies (sensors, command-and-control and fire systems) first emerged. However, these views were supported only by a select few and were not adopted by the wider Israeli defense establishment. The conventional wisdom posited that a transformation to defensive doctrine would signal to the enemy that the IDF had no intention to invade their territory. This, it was argued, would decrease the deterrence effect and the enemy would be disinclined to invest heavily in building defensive measures and reorient itself towards the offensive posture. Israeli decision makers assumed that even if based on a remarkable arsenal of precision guided munitions, defensive military doctrine would not be able to stop the enemy waves.

Although not accepted by the mainstream, these ideas did not go away; and during the 1980s, the "reformers" inside the IDF offered an alternative for the offensive breakthrough battles. Instead of breaking through an almost impenetrable Arab defense, they proposed exhausting enemy forces and inflicting heavy losses in the front and in the rear by using air force, navy, and special operations forces and by capitalizing on Israeli superiority in stand-off precision guided munitions, command-and-control systems and target-acquisition capabilities. They argued that by utilizing a qualitative edge in technology and human skills, and embarking on the defensive concept of operations, Israel could attain significant strategic benefits. Subsequent envelopment maneuvers against a weakened enemy would be far less impressive than the victory in 1967 but would minimize attrition rates. Their arguments were also supported by the recommendations of the report issued in 1987 by the Knesset Foreign and Security Affairs Committee (also known as the First Dan Merirod Report), which recommended paying additional attention to defensive form of warfare, both in terms of the concept of operation and force build-up programs.

Additional imperatives pushed towards further transformation of the IDF doctrine when the confluence of several political, social, technological, and economic developments began to redefine the Israeli approach to military affairs in the 1990s, expanding the role of defense in the Israeli military modus operandi. Digesting lessons learned from the First Gulf War and the US IT-RMA ideas, the IDF gradually assimilated the idea that offensive mobility in its own was not an ultimate solution for the modern battlefield, and started to transform itself into a "small and smart military." These ideas, inspired by the IT-RMA concepts, strongly echoed the arguments of the "reformers" from the 1980s, who called for defensively oriented military doctrine. The reformers argued in favor of destroying the enemy deep inside its territory without crossing international borders and maneuvering precision fire in place of heavy forces. The overall influence of this idea on the Israeli defense establishment was gradual, but it introduced the first strong stream of defensive thought into Israeli security discourse and solidified the conceptual foundation for deterrence by denial.[43]

In addition, lessons from the First Gulf War resulted in the increased attention to the passive defense initiatives and projects. On the military side, the Home Front Command was established in the IDF in 1992 as an additional regional dimension of warfare to supplement the existing three regional commands. Inauguration of the new command symbolized the awareness that the rear turns into a battlefield in itself and that traditional regional commands are insufficient in providing military countermeasure to the threats on the hinterland. On the civilian side, the government issued a regulation for the construction industry to provide every new apartment with a mandatory protected space—a shelter popularly known in Hebrew as *mamad*. Since then, and especially following the Second Lebanon War, critical infrastructure and installations (both military and civilian) have been systematically strengthened and better defended to withstand even the hits from accurate rockets and missiles. This national-level endeavor of the total introduction of the *mamad* manifests a hidden assumption that neither preventive offensive military operations nor active defense can offer an effective response to the threat and that a passive, non-military defense is now part of the Israeli national security resilience.

Secondly, the waves of suicide terrorism and the growing *ballistic muqauma* since the 2000s pushed Israeli strategists to pay even more attention to both active and passive defense. The following three issues are particularly worth mentioning. The first was the Western Bank Barrier, also known as the Separation Fence, which has been under construction since 2000 to protect civilians from the Palestinian terrorist attacks emanating from the West Bank. According to Israeli estimates, the functioning segments of the fence decreased the number of successful penetrations into Israel. The work on the separation fence has continued during the last decade, aimed to hermetically envelope Israeli borders with the Palestinian Authority. Disengagement from Gaza was also in a way inspired by this notion of the Separation Fences' effectiveness. The fence was supplemented by the increased and sophisticated defensive measures inside the Israeli hinterland, such as checkpoints, guards, and metal detectors at buildings' entrances.

The second issue relates to the importance that Israeli strategic discourse began to attribute during the last decade to the defensive approach in national security. In the spring of 2006, several months prior to the Second Lebanon War, the committee charged with formulating the Israeli National Security Concept, also known as the Dan Meridor Committee, submitted its final report to the prime minister and the minister of defense. The report that was prepared for two years explored, among other issues, the changing character of war and argued that in the next round of warfare the "civilian rear" will ultimately turn into a "military front" and thus will be the main zone of the future combat activity. The committee recommended to the government to add to the traditional Israeli national security triangle the "fourth pillar"—defense. This new pillar of national security encapsulated a greater emphasis on the defensive form of warfare and introduction of the bulk of non-military, passive defense activities in the civilian realm. It is worth noting that in the professional jargon of Israeli experts the Hebrew term for this additional pillar is often not *hagana* (defense) but *hitgonenut* (the term for defense used in the non-military context; probably the closest approximation to civil defense in English). Defense (*hagana*), according to the classical Clausewitzian logic and traditional Israeli military thought, is defined as a form of warfare that should prepare necessary conditions for the subsequent offence. Here, on the contrary, it refers to the stand-alone passive and active protection of the civilian rear and is not perceived as a precondition for the subsequent offensive operation.[44]

The reality of the Second Lebanon War and the subsequent Winograd Report underscored these recommendations of the Meridor Committee. The "defensive narrative" penetrated not only the discourse but also the national security practice. The National Emergency Authority (*Reshut*

Heyrum Le'umit) was created in September 2007, charged with coordinating military and civilian activities during a state of emergency, war, or natural disaster. The Home Front Defense Ministry—an umbrella (somewhat redundant) organization charged with coordinating all active and passive defense efforts in Israel—was established in 2011. This move entailed an effort to build a new command and control system, separate from the IDF and the Israeli MoD and to transform it to an organization that is not responsible for winning the war.[45]

However, probably the main notion that Meridor Committee injected into the professional discourse among Israel experts and commentators was the following: traditional formula of the offensive operations (punishment) aimed for recharging the batteries of deterrence, should be supplemented and coordinated with defensive endeavors (denial). The latter can also make a meaningful contribution to the maintenance of stable deterrence regime and postpone the next round of violence.

Finally, the third issue to boost the importance of deterrence by denial refers to the missile defense systems. The tendency started following the 1991 war, when for the first time in Israeli military history significant funds were allocated to research, development and acquisition of the pure defensive weapons' system—Arrow Ballistic Missile Defense Project. (Barrack anti-ship missile project is another significant investment in the weapon system that occurred around the same time.) This inclination further gathered momentum during the Second Intifada and eventually matured into one of the main trends of the national security policy after the Second Lebanon War. In line with the Meridor Committee's recommendation, but for the reasons primarily related to the course of the Second Lebanon War and public critique after it, active defensive capabilities received stronger attention and financial support from the government. The Iron Dome missile defense batteries were rapidly developed and procured, funds were allocated for further R&D in this field, and several civil defense organizations and doctrines were inaugurated. In the future the Iron Dome (short-range and lowest-tier) interceptor, together with the Arrow (long-range) and the David Sling, also known as Magic Wand, (mid-range) missiles interceptors, are expected to provide a layered missile defense over the Israeli territory and to counter ballistic threats from various ranges and locations. This interceptors' system of systems should allocate the right interceptor to the right target and respond to all the threats simultaneously.

The vision of the future conflict further enhanced the role of Israeli missile defense and consequently resulted in the greater role attributed to deterrence by denial. The mainstream of the Israeli experts imagines the future battlefield as a ballistic strikes' storm from Iran, Syria, Hamas, Hezbollah, and possibly from the West Bank, and from the ungoverned territories on the Israeli borders (Sinai and the Syrian Golan Heights). Traditional Israeli offensive standoff fires and ground maneuvers are not always possible and effective against the threat of several thousand missiles and rockets launched against Israeli civilian areas from various civilian locations outside Israel. Thus, Israeli response to this counter-value ballistic threat is triple—prevention, active defense, and passive defense – and its deterrence posture is more balanced between punishment and denial. This juxtaposes Israeli's coercion approach to missile defense from Japan's, which, as Trexel illustrates in his chapter, largely developed in the "absence of offensive kinetic military cost-imposition capabilities." The introduction of the Iron Dome system contributed to the feeling of security among the civilian population. It also became appealing to policy makers, as a relatively effective alternative to more controversial and operationally complicated offensive ground maneuvers. As such, missile defense expanded the options of the politicians not only in terms of the military arsenal but also on the diplomatic front, especially in the realm of the battle for legitimacy. In contrast to the traditional Israeli view of defense, here it was not seen either as a precondition for the subsequent offense, or as the way of preserving operational achievements produced by the offense, but as a solution in itself. Moreover, this capability enabled Israeli leadership to construct and benefit from the international legitimacy and to wage a campaign for world public opinion—a sphere to which Israeli decision makers started to pay more attention during the last decade. In addition, the legitimacy gained at the initial stages of the campaign by using defensive capabilities can be, in theory, then utilized as justification for the subsequent offensive action if the fires from the other side continue. Currently, the Israeli strategic community is seeking a balance between these three efforts and investigating the amount of funds to allocate to each. The three efforts should be simultaneous and mutually reinforcing, so that offensive and defensive endeavors complement one another. Offensive capabilities should reduce launching capacity, so that no more than 20 percent of the adversaries' ballistic arsenal can hit the Israeli hinterland, including the areas of metropolitan Tel-Aviv. This percentage of strikes is tolerable for active and passive defense.[46]

Prevention entails *mabam* and offensive operations that should destroy the launchers before or right after the first launch. Theoretically, this real-time prevention will be conducted by precision-guided munitions, based on real-time intelligence and invasions by ground forces. However, under the changing character of war, the time that is needed for preparing and launching a ballistic strike is much shorter than preparations that Israeli enemies once conducted for the ground invasion. Given this short preparation time, the window for the preventive offensive strike is shortened and thus defense can offer a better operational solution. In addition, from a legitimacy perspective, the offensive option, by definition, is more problematic than a defensive one because the enemy is purposefully operating from within civilian areas.

Although missile defense will probably not affect the motivation of the other side, it can dissuade it from initiating another round of violence by communicating that future offence may be futile in light of the Israeli denial options and costly, given the possibility of the subsequent Israeli punishment. The lowered expectations from the potential strike may force the other side to reconsider its execution. Taken together these recent trends produced a much stronger emphasis on defense in Israeli national security and thus somewhat expanded the role of deterrence by denial in Israeli strategic thought. Incrementally, deterrence by denial began to equalize its role when compared with deterrence by punishment achieved through offensive operations.

CONCLUSION

This chapter focused on the Israeli approach to deterrence and specifically examined the role of deterrence by denial in the Israeli strategic thinking. Deterrence has been one of the three main pillars of the Israeli national security concept since the late 1940s. From the onset, the Israeli concept of deterrence had greatly differed from the Western nuclear conceptualization of deterrence and resonated more with the less developed part of the IR deterrence research program—a conventional one. Deterrence has been a by-product of ultimate military victory that the IDF was expected to produce in each round of military clashes vis-à-vis state enemies, or an outcome of the retaliation operations against non-state foes. Thus, in its essence, historically, it was first and foremost deterrence by punishment, not deterrence by denial that became the default option of the Israeli approach.

Since the 1990s, against the backdrop of the changing character of war, the traditional deterrence concept became obsolete and the Israeli strategic community began to experience a knowledge crisis. The latter led to the paradigm shift in Israeli strategic thought and made periodic deterrence operations and an ongoing campaign between the wars into the main generator of the deterrence regime. Israeli practitioners questioned whether deterrence stems from battlefield decision and realized that the changing character of war had incapacitated the IDF from generating a battlefield decision. Both insights further diminished the role of traditional offensive operations aimed at battlefield decision and subsequently left more room for defense and deterrence by denial. This shift in deterrence paradigm coincided with several defensive capabilities and defense-related activities, which strategic studies traditionally associate with deterrence by denial, and which started to gather momentum in the Israeli approach.

The last decade manifested the strongest ever inclination of the Israeli strategists toward defensive mode of operations and consequently, toward deterrence by denial. First, the lessons learned from the Yom Kippur War and the First Gulf War reoriented Israeli military thought, shifting it away from an overall emphasis on offensive mobility and toward a greater emphasis on defense in terms of operations, organizational structures, weapons systems, R&D, and procurement. New weapon systems such as Arrow and Barack missiles, the establishment of the Home Front Command, and turning each apartment into a personal shelter demonstrated that already in the 1990s offensive mode of operations left to the defensive measures significant place in national security policy. This tendency was reinforced during the 2000s when the Separation Fence, hardening of civilian targets, and anti-missile defense systems became significant countermeasures against the asymmetrical enemy that tried to "design around" the traditional Israeli offensive approach. This stream of defensive thought and action was codified and institutionalized by the Meridor Committee Report that proclaimed defense as an additional pillar of the Israeli national security concept.

Consequently, the traditional formula of offensive operations (punishment), aimed at recharging the batteries of deterrence, was supplemented and coordinated with defensive endeavors (denial). The latter is expected to contribute to the maintenance of stable deterrence regime and postpone the next round of violence. Defense and deterrence by denial afforded Israel additional strategic options, beyond those following the scripts of Hamas and Hezbollah to punish and preempt invariably and continuously. They provided the population with a feeling of security, ensured legitimacy of the Israeli actions in the eyes of the international community, and granted Israeli decision makers greater space for military and political options.

However, despite the fact that deterrence by denial acquired unprecedented importance and took over several missions that were previously in the jurisdiction of deterrence by punishment, and despite the most important role that the active missile defense played in Operation Pillar of Defense, most Israeli strategists do not consider capabilities associated with deterrence by denial to be more effective than offensive "deterrence operations" in preventing an outbreak of future violence. Why are Israeli practitioners still somewhat skeptical about the potential of defense and, subsequently, the potential of deterrence by denial?

As far as passive defense is concerned, the mainstream assumption is that it was not the Separation Fence but rather offensive combat activity, such as Operation Defensive Shield, that deterred further acts of violence by the Palestinians and eventually led to the Israelis to prevail in the Second Intifada. Another widespread opinion suggests that despite its operational effectiveness, the Separation Fence did not decrease the motivation for *muqauma* but only channeled the operational violence into an alternative, ballistic realm. For this reason, Israeli strategists do not see the Separation Fence as the ultimate tool of effective deterrence, although one can indeed argue that it constitutes an example of effective deterrence by denial in action.

Without a doubt, the Meridor Report added value to the deterrence-by-denial strategy by introducing a "defensive narrative" into the Israeli strategic mindset that was traditionally strongly dominated by the cult of the offensive. However, many of the recommendations of the Meridor Committee remained as general strategic declarations and were neither approved as government policy nor translated into concrete directives.

As far as active defense is concerned, the boost of missile defense systems did not fundamentally change Israeli approach to deterrence. The last "deterrence operation" demonstrated that effective functioning of the Iron Dome contributed to the IDF's decision to refrain from ground offensive. However, even if the Israeli strategists viewed the Iron Dome as having made the civilian population feel secure and shifted the traditional offense-defense balance of the IDF's approach, the primacy of "deterrence by punishment" still stayed intact. The Israeli strategic community consists of competing schools of thought on this matter. Whereas some see defense as a minimal component of military policy, others argue for a more balanced proportion. Nevertheless, the offensive school of thought continues to prevail.

One set of considerations is strategic-operational. Some Israeli experts do not see missile defense as a game changer in the punishment-denial balance. According to them, even effectively functioning Iron Dome and other MD systems are only a partial response to the perceived future rocket-missile rain. Israel's enemies are expected to further improve the range, accuracy, payloads, and numbers of their projectiles. Experts tend to overlook the ratio of potential damage produced by a successful ballistic hit and the price of interception. Instead, the tendency is to focus on investments, not outcomes. Hence, it is widely argued that the ratio of production prices and paces for missiles and interceptors are enormously different—tremendously unfavorable for the active defense. Thus, in the observable future, Israeli enemies may be able to overwhelm and outflank these missile defense systems, not to mention that full coverage may simply be unaffordable for Israel.[47] Because the hermetic coverage is impossible for financial and operational reasons, it is unclear what the defensive endeavor should be focused on. Should the government protect civilian installations, or instead cover military targets, or, alternatively, defend forces operating on the ground?[48] In this sense, the Iron Dome success story is not different from the post-1991 Israeli decision to advance with the Arrow system—a major project that has matured but is still a secondary effort when compared to offensive capabilities.

Moreover, during and after Operation Pillar of Defense, some non-Israeli and, in particular, Arab commentators asserted that the IDF's decision to refrain from ground maneuver was not based on the Iron Dome but rather from a lack of resolve and insufficient offensive capability on the part of the IDF. Some Hamas commentators described the IDF's refrain from invasion as successful Hamas deterrence and victory.[49] This assertion is obviously intolerable to IDF commanders because it immediately and directly diminishes the IDF's image they are trying to maintain of being a force of deterrence. Thus, even if missile defense enables the IDF to avoid some difficult choices (e.g., ground maneuver), the basic instinct still pivots toward deterrence by punishment.

It also appears that the IDF still prefers striking its enemy's ballistic arsenals at the "phase zero" (pre-combat) phase[50] and waging *mabam* aimed at downgrading arsenals at the procurement stage. This would be the way to dissuade enemies from the start of the campaign—a clear preference of punishment-over-denial alternative. Military force build-up visions and programs for the next decade clearly demonstrate that the IDF is increasingly paying attention to defensive capabilities. However, most of the budget allocation and conceptual emphasis in all the services go to offensive precision fire, maneuver, and cyber operations, as well as to intelligence capabilities supporting this type of warfare. As long as there is no fundamental technological-conceptual breakthrough that proves the effectiveness of the active defense systems, defense and deterrence by denial are likely to continue playing an axillary and relatively minor role in the Israeli strategy.[51]

Finally, from the international legitimacy perspective, rather paradoxically, missile defense may turn into a double edge sword. Some Israeli experts argue, that if active defense systems prove themselves effective, then the international community will expect Israel to rely on them alone and will not tolerate Israeli offensive operations.[52] The other reason for reluctance is essentially psychological and is related to Israeli strategic culture. In keeping with the Israeli culture of war, offense preserved its primacy over defense both in the mindsets of the strategists and in the actual operational planning. Israeli strategic mentality is designed to address perceived challenges and existential threats by the pro-active, offensive means. One is expected to take care of the pressing national security challenges through an active prevention or offensive response, and not through hiding behind fences or other protective measures. The latter approach resonates with the historical conduct of the fearful and oppressed diaspora Jews, who were not able to take their fate into the hands, and thus runs against the Zionist ethos of the people of the plow and rifle. Moreover, the offensive approach matches well the culture of improvisation, stratagems and tactical ingenuity—long lasting traits of the IDF military tradition. Consequently, the role of deterrence by denial may be marginalized, as long as this trait of the Israeli strategic culture stays intact.

Paradoxically, despite this view of the Israeli strategists, actions taken in the denial (and defense) realm might have a strong deterrence effect on Israeli adversaries. However, in order to establish this point and to prove the effectiveness of "deterrence by denial," a separate inquiry into the strategic calculus of Israeli adversaries is due. Such inquiry may further persuade the Israeli strategic community in the effectiveness of deterrence by denial. This effort is however beyond the scope of this chapter.

Notes

1. See chapters 1, 2, and 5.
2. Rid, "Deterrence Beyond the State," 124–147.
3. Rid, "Deterrence Beyond the State," 125. For example, see Lebow, "Thucydides and Deterrence," , 163–188; R. J. Overy, "Air Power and the Origins of Deterrence Theory"; Quester, *Deterrence Before Hiroshima.*
4. This resonance is accidental and inadvertent because the theory of law has not been a source of inspiration for Israeli strategists. However, viewed from the lens of law theory, the Israeli approach resembles a criminological one. See chapter 3 in this book.
5. Bar-Joseph, "Variations on a Theme," 146.
6. Rid, "Deterrence Beyond the State," 129
7. See chapter 6.
8. "Israel's experience illuminates the relationship between the deterring use of force and the construction of norms." Rid, "Deterrence Beyond the State," 125.
9. Lieberman, *Reconceptualizing Deterrence*, 220; Rid, "Deterrence Beyond the State," 141.
10. Lieberman, *Reconceptualizing Deterrence*, 23. See Mearsheimer, "Prospects for Conventional Deterrence in Europe," 158–162.
11. Lieberman, *Reconceptualizing Deterrence*, 1–43. Even if conventional war approaches its totality, "leaders could still hope to find a strategy that would provide them with victory." Lieberman, 20.
12. Michael Gerson, "Conventional Deterrence in the Second Nuclear Age," 42. Lieberman, *Reconceptualizing Deterrence*, 9–13.
13. Interview with senior (ret.) IDF officer, July 3, 2012; Brom, "Israel's Missile Defense."
14. Such as general versus immediate; broad versus narrow; existential versus extended. Paul, Morgan, and Wirtz, *Complex Deterrence*; Freedman, *Deterrence*; Morgan, *Deterrence*, 28–43.
15. Lieberman, *Reconceptualizing Deterrence*, 213. For the detailed analysis and discussion, see Gross Stein, "Deterrence and Learning in an Enduring Rivalry."
16. Bar-Joseph, "Variations on a Theme," 148
17. Bar-Joseph, "Variations on a Theme," 145–181. See also Almog, "Cumulative Deterrence and the War on Terrorism," 4–19.

18. Atzili and Pearlman, "Triadic Deterrence"; Pearlman and Atzili, *Triadic Coercion.*

19. See sources in Adamsky, *The Culture of Military Innovation,* 112–113.

20. Lieberman, "Deterrence Theory."

21. Almog, "Cumulative Deterrence and the War on Terrorism."

22. Adamsky, *The Culture of Military Innovation.*

23. Handel, "The Evolution of Israeli Strategy"; Maoz, *Defending the Holy Land.*

24. Maoz, *Defending the Holy Land.*

25. Lieberman, "Deterrence Theory," 22

26. See Amir Lupovici, "Cyber Warfare and Deterrence."

27. Maoz, *Defending the Holy Land,* 231–301.

28. Henriksen, "Deterrence by Default?"; Feldman, "Deterrence and the Israeli-Hezbollah War," 281.

29. *Deterrence* aims at maintaining *status quo*; *compellence* aims at its restoration. George, "Coercive Diplomacy," 71. To introduce new rules of the game, proportional force does not suffice.

30. Lieberman, "Deterrence Theory," 219.

31. For example, see Bar-Joseph, "Variations on a Theme".; Lieberman, "Deterrence Theory."

32. The term is borrowed from George and Smoke, *Deterrence in American Foreign Policy.*

33. The term is borrowed from Rosen, *Winning the Next War.*

34. Brun, "The Other Revolution in Military Affairs."

35. Slater, "Just War Model Philosophy."

36. For example, see Eran, "Israel and Weak Neighboring States," 6–8; Petrelli, "Deterring Insurgents."

37. "Takhlifei Hakhraa," in Hebrew. Interview with senior defense official, August 7, 2012 (Jerusalem); interview with senior IDF military officer (ret.), July 3, 2012 (Tel Aviv).

38. Interview with senior IDF military officer, July 1, 2012 (Central Israel).

39. For example, see Amos Yadlin, "Time for Decisions," 78–79.

40. Rid, "Deterrence Beyond the State," 127–128; See chapter 3 in this volume.

41. Hendel, "Shuvo shel maarach hamiluim," 32.

42. Oren, "Et hadegel al hagiva itkeu hametosim"; Rapoport, "Hatkifa bein hamilhamot"; Shuki Tausig, "Mabam," *Ha'ain Hashviit,* February 1, 2013; Or Heler, "Syria rotza hishbon," *Globs,* February 4, 2013.

43. Adamsky, *The Culture of Military Innovation.*

44. Interviews with senior IDF officers, Spring 2014.

45. Amidror, "Missile Defense."

46. Uzi Rubin, "Israel's Missile Defense"; Pedatzur, "How Missile Defense Undermines Deterrence."

47. Peter Dombrowski, "Demystifying the Iron Dome."

48. Amidror, "Missile Defense."

49. Dombrowski, "Demystifying the Iron Dome."

50. Ibid.

51. Harel, "Kah terae milhama be 2025."

52. For example, see the discussion in Stenzler-Koblentz, "Iron Dome's Impact on the Military and Political Agenda."

CHAPTER 8

CAN DENIAL DETER
IN CYBERSPACE?

Martin Libicki

Good defenses can limit the damage from cyberattacks and facilitate recovery. They permit states to threaten cyberattacks with less fear of retaliation. The more hopeful case for good defenses is that they discourage others from attacking in the first place (sometimes known as deterrence by denial). But do they? The answer is complicated. First, it depends on the type of cyberattack at issue. Second, it depends whether it is the cyberattack being defeated or it is the effect that the cyberattack is trying to produce which is being denied. The argument is that the deterrence effect of defense is difficult to induce and discern, particularly in comparison to a contest in more traditional media. However, there seems to be more deterrence potential in *denying* to the attacker the effects being produced by the cyberattack. These principles apply across the gamut of cyberattacks. They apply with particular force to high-end cyberattacks carried out by one country against another country either against the latter's critical infrastructure (strategic cyberwar) or against its fielded forces and their support systems (operational cyberwar).

This chapter first considers the direct deterrent effects of denial by discussing the distinction between discouraging investment in cyberattack capabilities versus discouraging the use of such capabilities: it is often difficult to predict how well cyberattacks may succeed until one actually invests in such capabilities to carry out a probe of the target system (which generally constitute most of the capabilities needed for an attack). The analysis focusses on canonical cyberattacks (those that work from penetration to subversion) but note that there are other cyberattacks (e.g., DDOS attacks). It then shifts from considering the attacker as a rational unitary actor to incorporate psychological issues and the relationship between hackers and their leaders. The latter half of the chapter pulls the lens back to examine the strategy of discouraging cyberattacks by defeating the attacker's strategy, be it coercion or using cyberattacks to facilitate follow-up kinetic attacks.

DIRECT DETERRENCE BY DENIAL

Consider conventional warfare undertaken for conventional purposes: to seize land, overthrow a government, or shape the psychology of future relations.[1] As several chapters (e.g., by Patrick Morgan, James Wirtz, Jonathan Trexel, Dima Adamsky) in this volume illustrate, in contemplating whether to initiate conflict, a rational state compares benefits to costs.[2] The ability to mount an offense is an aspect of such costs[3] measured in terms of effort and lost assets. Such calculations depend critically on the relationship between the resources that must be invested to mount an attack and the resources that are expended to carry out an attack. If the latter are low, then having already decided to make the necessary investments, the decision to use them is likely to be influenced by the prospect of and losses from retaliation—but if the expected costs from retaliation are themselves low, the prospect of failure is likely to be of but modest importance because mounting the attack costs little. In such a case, it is the decision to invest rather than the decision to expend resources which is what will be influenced by the prospects of success. Normally, the same kind of information (on the adversary's capabilities) that helps in judging whether an attack can achieve its goals also goes into judging whether investing in an attack capability can achieve its goals. But that equation falls apart if the investment required to determine the odds of success is a large fraction of the investment required to achieve success. That understood, the dim prospects for deterrence by denial can now be demonstrated.

CANONICAL CYBERATTACKS

This analysis starts with assuming a canonical cyberattack.[4] Such an attack starts with general surveillance of the target, followed by the penetration of the target for the dual purposes of specific surveillance *and* the insertion of malware. This malware, in turn, accepts the attacker's commands and has the infected computer run them. Such commands constitute the means of attack—the weapon as it were.

An effort made to attack in cyberspace may be divided into three parts. One part is general investment; it gives the attacker capabilities against all targets or at least all targets of a particular class (e.g., electric power grids). A second part is specific investment: the search for particular vulnerabilities, standby mechanisms, and an understanding of the relationship between information and operations in the target's various systems and commands. A third part is composed of specific operations— the time and effort required to carry out, monitor, and provide feedback on an attack of particular targets plus whatever assets are used up in the process of carrying out such operations. If the attack exploits a unique vulnerability or contains a unique exploit—in such a way that discovery of the exploit prevents its re-use against this, or worse, any other target— then the likelihood of its loss has to be included in the cost of carrying out an attack. Similarly, if the attack uses a set of tools (e.g., tricks for evading detection) which are then exposed, the effort to acquire substitute tools has to be considered part of that cost. In general, if a large percentage of the effort goes into collecting intelligence on the specific target, then the decision to undertake such an effort depends on the likelihood that such an investment will pay off. Unfortunately, the attackers may have to make target-specific investments to discover that the target's defenses are daunting. If, at that point, the extra costs of actually carrying out an attack are modest, the attacker may feel that it has little (apart from revealing the target's vulnerabilities to itself) to lose by trying, even if the odds of success are low. If all else fails, hackers will have received live-fire training.

Assume now a target with perfect defenses. The attacker attempts a series of attacks on one target, gains nothing every time, concludes that it faces no good prospects of success, and decides not to waste its resources trying to attack that target in the future. Here, defense discourages, but is discouraging such attackers worth anything to the target itself? If, having invested in defense (e.g., a fortress wall), the defenders make no further effort beyond the routine monitoring (e.g., patrols along the fortress wall) that it would have made whether or not an attack is in progress, then it matters little what effect their defensive efforts had on what attackers decided to do. In other words, the benefits of deterrence do not matter to the defenders. Indeed, the target is better off having adversaries waste their efforts if the alternative is such adversaries redirecting resources into something more dangerous. Such logic does not apply to violent conflict where money and blood will have been spilled even if victory is complete; it is worthwhile to keep the enemy from starting a fight —even a fight they lose.

In practice, exactly how costly a failed attempt is to the defender depends on how deep the attackers got. Imagine an attack in which multiple client machines (e.g., personal computers) within the organization were infected in an attempt to penetrate an enclave within the network (e.g., the machines were part of the office automation system whereas the enclave contained controls for the machinery that extracted oil and moved it around). The enclave withstands a breach; the attack has failed in its primary objective. However, this leaves the network's system administrators with the unenviable task of discovering all the infected machines and cleaning them up (lest the infected machines be used as a jumping-off point for a next attempt). If the attacker could have been persuaded that it could not breach the enclave, then it might well have avoided infecting the client machines in the first place, thereby sparing the target the cost of cleaning up the infection.

Some forms of defense—in this case, deception[5]—are particularly weak in providing cyber deterrence. Normally, deception is used to convince attackers that they have succeeded when they have failed: for example, that what looks like the mother lode of intelligence turns out to be fool's gold, or the lever that they now control actually connects to nothing. Making attackers think they have succeeded is likely to spur them on. Only later when they realize they have been tricked might they be discouraged from further attempts. In the interim, they are not discouraged hence not deterred.

The Difficulty of Convincing Attackers that Defenses are Good

This example now raises the difficulty: how does the attacker know that its efforts are doomed to failure? How does the target convey as much to the attacker? In historical physical combat, where the nature and, often, the quantity of the other side's weaponry are visible—as both Wirtz and Adamsky illustrate in their chapters—the potential attacker can guess the chances of success.[6] It may convince itself that it will win because it possesses immeasurable attributes such as superior élan or generalship, but even self-deception has its limits. The more manpower and equipment the defender can show to the world, the less confident the attacker will be of winning. The advent of electronic warfare and the rising importance of having tactical intelligence on the other side (e.g., to find where they are hiding their assets) suggest the growing uncertainty of such calculations for even kinetic warfare, though. An attacker might understand what tricks it has but not how well the other side can thwart such tricks or generate tricks on its own.

In cyberwar, where a vulnerability discovered is a vulnerability neutralized, the difficulty of convincing others up front that your systems will not fall for their tricks is more complicated. In essence, an attacker can take down a well-tended system only by knowing something about the system (e.g., a vulnerability) that the defender does not know. The best a defender can do is to demonstrate its ability to defend the system against a set of known class of attacks and hope that the class is sufficiently encompassing that what remains is of little practical importance or else presents much higher risks to the attacker (e.g., it requires using special forces to penetrate physical barriers). More broadly the target may have a reputation for defeating the wiles of others. Otherwise, the only way the attacker is going to really know the impenetrability of a target is by trying to penetrate it. If the effort succeeds, the attack is on. Alternatively, if there are multiple barriers, then the same problem arises when trying to surmount the next one (only worse—if there is less public knowledge about the difficulty of surmounting deeper barriers vis-à-vis surficial ones). If the effort fails, the attacker has to calculate the likelihood that, with N failures, the N-plus-1st effort will be a success. It may use Bayesian logic to convince itself to give up,[7] or it may reason that having invested so much effort to learn what would not work, the odds of finding what would work on the next attempt rise with every failure.[8] The relationship between past failure and future prospects is not straightforward: initial defeat may not necessarily deter later attempts.

If an attacker tried to repeatedly penetrate multiple systems over time (typical for cyber espionage), it might foresee ultimate failure through, say, successively smaller harvests of information.[9] But it may, instead, save cyberattacks for when they might do the most good because it fears that wasted attacks may reveal vulnerabilities and stiffen the target's defenses. If so, its opportunity to detect a good defense may be limited.

At best, it can detect increasing difficulties in penetrating systems when trying to spy on them and thereby conclude that cyberattacks would also become more difficult. However, if the target's defenses are deeper than simple anti-penetration devices (e.g., better backup, adroit monitoring, greater overall resiliency), the quality of these defenses may be unseen and therefore irrelevant for their never having been invoked.

The Fuzzy Relationship between the Prospects of Failure and the Effort in Trying

The correspondence between the prospect of failure and effort devoted to trying is even fuzzier if the investment required to penetrate a system has very little to do with the specific system itself. A great deal of effort on the part of hackers is entailed in finding exploitable vulnerabilities in software (e.g., Adobe Acrobat) used by a large number of potential targets. The attacker may have a hundred comparable targets; perhaps it is interested in taking down a bank's computer as a psychological operation and the target country has multiple banks. Even if any target bank decides to mount a perfect defense, this hardly discourages the attack. The attacker's prospective gain from investment goes down by one part in one hundred. If attacks are effortless once the investments are made (putting aside the fact that monitoring targets over time is part of target-specific investment), the attacker has no reason not to attack such a well-defended target. Indeed, it is hardly worthwhile differentiating the hard targets from the soft targets. Just attack them all.[10] If this sounds bizarre or unusual, it is a fair characterization of the effort required to recruit bots for botnets. Bot-herders generally spend their resources developing or acquiring vulnerabilities and then distributing their malware, such as bad PDF files or corrupted web sites, without much regard to who may pick it up. A similar attack is called the watering-hole attack: malware is inserted in a popular website (often by infecting an advertisement hosted, not by the website owner, but by something renting the space from the website owner); all who pull down the page absorb the malware into their system. Some of those absorbing the malware are consequently infected.

Although the mass use of safe computing practices may discourage such efforts, the efforts of any one user will not discourage attackers but merely limit or eliminate the attack's effects on that user. Herein lies a difference between discouraging *investment* in developing cyberwar capabilities and discouraging the *use* of such capabilities.

Overall, the greater the role of "generic investment" (e.g., looking for zero-day vulnerabilities rather than target-specific investments, such as understanding specific failure modes) in building cyberattack capabilities, the less discouragement a good defense will provide to potential attackers. As a general rule, the softer the target set, the more effort will be devoted to general investment in broad-scale attack tools and the less attention will be paid to the particular defenses of any one target. Conversely, the harder the target set, the more that its vulnerabilities will be unique (intelligence agencies can spend months scoping a high-value target waiting for a flaw to present itself) and the more attention will be paid to target-specific operations. In either case, however, the relationship between initial discouragement and ultimate withdrawal is uncertain.

It merits noting that the mix of investment and operations has not been a constant over time. In the late 1990s, hacking was more typically done by hackers—individuals skilled at getting into systems and trying this and that movement while defenders were hastily erecting barriers to their virtual movement.[11] Such activity happened in real time, providing hackers rapid feedback on whether their efforts were succeeding and thus worthy of continuation. These days, more effort is involved in building and deploying tools, which means that most of the commitment precedes contact. If initiatives such as DARPA's self-healing networks work and are employed broadly enough, real-time human intelligence operating in real-time may again be required to penetrate computers.

OTHER FORMS OF COMPUTER ATTACK

Although the cyberattacks of greatest note involve the penetration of target computers, they are not the only form of cyberattacks, and any relationship between denial and dissuasion depends on what form of cyberattack is being considered.

One class of cyberattacks, often used to support the cybercrimes associated with identity theft, is called SQL-I (structured query language injection). The attacker, acting as a random user, interacts with the target system by feeding it legitimate-looking commands to trick it into revealing information on others rather than just about himself (e.g., a user asking: what is my balance?). One such attacker walked away with 3.6 million tax records from the State of South Carolina.[12] As a rule, to carry out such attacks, hackers try a variety of approaches until one works. Thus denial, after a while, might dissuade.

Another class of attack entails infecting individual computers (or more recently, servers) and turning them into bots. These bots are then used to drive so much traffic into the target that the target is knocked offline. This is a two-level attack: first recruiting the bots, and second turning them on a target. Unless the attackers have a very specific interest in only one particular target, the hardness of any one recruitment target is irrelevant to the decision to build the botnet. However, the hardness of the target—measured either in terms of its bandwidth[13] or its ability to withstand special tricks[14] may discourage attacks on it, leaving the hackers to direct their botnets elsewhere.

PSYCHOLOGICAL FACTORS

The opacity and ambiguity of cyberwar suggests that even the consequences of a perfect defense may not be clear-cut to potential attackers. The attacker's decisionmakers will have little direct knowledge of whether attacks by third-party attackers on the purportedly well-defended target are succeeding; all it knows for certain is that there is no visible success yet. Whether their own cyberattacks succeed is knowledge that, for the most part, only the cyberwarriors possess.

If the hackers persuade their bosses that the right goals—for example, hindering the target's ability to make decisions—have been met but are hard to measure, who would know they failed? The hackers themselves may be discouraged, but, if the *raison d'etre* of the cyberwar bureaucracy were at risk from the delivery of bad news, they may hold their tongues. Without bad news, the attacker's decisionmakers have no way of knowing that their investment is futile. Thus, they are not necessarily discouraged from trying again.

Plausible irrationality may also color the effects that good defenses can have on the willingness to attack. A reasonable attacker may presume that, after so many tries and no successes, the prospects of further success are dim. But it is human to believe that the fault may be not in the difficulty of the target but in the failure to make adequate effort.

The more people invest in a problem, the more likely they are to press ahead and try to recoup their losses—the certainty that people can recognize and walk away from sunk costs as such is a conceit of economists, not psychologists. The dynamic nature of cyberspace can convince one that targets that seem impregnable today may be vulnerable tomorrow simply because things change all the time, so keep trying.

Several other considerations merit note. Economic theory says that the greater the price of something, the less people will want it: If the price of potatoes rises, people will eat pasta. If the price of success in cyberwar is high, people will find other ways of hurting their enemies. But the size of the relationship depends on the elasticity of demand. A state committed to achieving an effect, and finding it harder but not impossible to do, may elect to throw more resources at trying.

Conversely, even an imperfect defense may persuade attackers to stop cyberattacks altogether. An attacker may reason that an attempt to harm a computer via cyberattacks will lead to the discovery of the attack and may lead the target to discover how the initial penetration was made. If such penetration techniques are discovered, then cyberespionage—which often also requires such penetration—becomes that much harder. The attacker may well refrain from cyberattacks to maintain its cyberespionage capability.

None of this says that defenses cannot dissuade, but such claims need to be viewed carefully and cautiously.

DISCOURAGING STRATEGIC ATTACK

Cyberattacks are rarely ends in themselves. They cannot seize assets (they do not seize information so much as copy it; the original is still there). Even a nasty cyberattack lacks the heft to oust governments. They cannot really disarm those they strike (unless the latter's systems are rigged to destroy themselves if given bad commands). That does not mean that cyberwar is useless—but its usefulness is instrumental rather than decisive on its own. A sufficiently bold attack on civilian systems may induce caution and respect on the part of a target state (or organization); this can be termed strategic[15] cyberwar. Similarly, a sufficiently clever attack on military forces can create a window of opportunity during which military operations have a greater chance to succeed. This may be termed operational cyberwar.

The instrumentality of strategic cyberwar suggests that it can be discouraged by making a credible case to any attacker that the psychological benefit it seeks will not be forthcoming. Thus, if a cyberattack is meant to keep its target country from intervening somewhere, for instance, the target can signal that it will stick to its policies regardless. If the attacker carries on, anyhow, it must necessarily hope that political forces (e.g., popular fear of a repeat) may bend the target state's initial posture.

Ironically, the attitude required to reduce the attacker's expectations that its attack will change the target's policy are almost the opposite of what conventional strategic thinking would call for. A narrative that says that cyberattacks on, for instance, a country's infrastructure are beneath notice is a very different narrative than one that elevates the importance of cyberattacks to the point that being the target of an attack can justify a kinetic response. The latter says that an attack has touched a nerve. A strategy of insouciance says the exact opposite. Similarly, policies to hype the threat, the better to make the owners of systems invest in protecting themselves against cyberattack are inconsistent with a posture of downplaying the effects of a cyberattack so as to render them strategically meaningless.

Such irony is not limited to cyberattack; it applies to terrorism as well. Hence the repeated advice of Bruce Schneier, a cybersecurity expert—refuse to be terrorized.[16] It is hard to argue against the proposition that the vigorous US response to the September 11th attacks helped raise the profile of violent resistance to the West in general and al Qaeda in particular—which may have been exactly what the terrorists sought in the first place.

Discouraging Operational Cyberattack

It may, however, be worthwhile to discourage operational cyberattacks undertaken to create opportunities to carry out kinetic attacks. What policies can a country follow to prevent the other side from acting on the hope that a sufficiently powerful cyberattack on its adversary's forces can disarm the latter long enough to achieve important goals?

For the United States, a commonly mooted example is a Chinese no-warning strike on logistics systems that paralyzes US force mobilization long enough to permit Chinese forces to secure a lodgment on Taiwan (or take the entire island). Against such possibilities, it is not enough to argue that good defenses against cyberattack will bring a foe's efforts to naught and will thus stop such a war before it starts. Such logic assumes that such a foe launches a cyberattack and measures how well the attack altered the correlation of forces—and goes ahead only if its cyberattack gives it sufficient advantage: a hack-check-fight sequence. The problem is that even if the results of the cyberattack are ultimately unsatisfactory, war could result, anyhow. Here are a few reasons:

- The leadership of the attacking state may not believe it can determine the results of the cyberattack reliably. It therefore commits to kinetic war beforehand, essentially betting that the hacking succeeds. Alternatively, they are psychologically primed to declare victory irrespective of actual results. After all, battle damage assessment, especially in cyberspace, is fuzzy under the best of circumstances. Proponents of cyberwar may want others to believe that their efforts worked and thus may proclaim success without fear of contradiction by non-hacker colleagues.

- Taking the time to assess the results of the cyberattack before deciding to use force may vitiate the element of surprise and therefore lead to less effective fighting. If the window of opportunity to take advantage of a successful cyberattack is measured in days (after which the targeted systems are restored and isolated/insulated from further mischief), the military must be mobilized before the cyberattack begins. Even if they do not act, their preparations for kinetic war may have to be so obvious by the time the cyberattack registers, that the attackers must start a fight or face a fight by the other side. Thus, the decision to carry out a cyberattack presupposes success and a follow-up by a kinetic attack whether the cyberattack succeeded.

- The attacking state may figure that if it starts with a major cyberattack the target state will respond with kinetic force whether the cyberattack was particularly damaging—and irrespective of whether the target can determine whether the attacker was mobilizing for a kinetic follow-up attack. The attacking state concludes that if any cyberattack leads to kinetic war, outcomes would be more favorable if it started on the offensive rather than the defensive.

So, the targeted country cannot necessarily rest assured that the ability to beat back a cyberattack will, in and of itself, keep it from facing an unwanted kinetic war. It needs to supplement after-the-fact capabilities to win such a war with some way to indicate to the other side that a first-hack-then-fight strategy will fail. This requires demonstrating either that hacking will fail on its own terms (i.e., the defender's systems are unaffected) and/or that even if hacking succeeds on its own terms (i.e., such systems are hurt), the target's residual ability to defend itself militarily will frustrate the attack.[17] Neither challenge is trivial.

How does one country convince others that its ability to conduct military operations is not at serious risk from a cyberattack? Successful cyberattacks generally require passing two sets of wickets. One set is getting into a system (considered a "set" because penetration may have to surmount multiple barriers). The other is making the system misbehave as the hacker wants. If the defenders are successful in keeping attackers out of their system, they start with the presumption of invulnerability. Unfortunately, consistent with the argument discussed earlier, attackers are likely to have a much better idea than the defenders (a.k.a. targets) of how well their penetration attempts worked. If the attackers cannot penetrate enough systems, they will likely judge that their hacking cannot succeed well enough to change the correlation of military forces appreciably. If they are informed and rational—and their efforts, in contrast to the previous discussion, have not been discovered—they will not continue plans to launch a kinetic attack.

The second set of wickets is the transition from penetration, to damage to defense systems (be it temporary or, better yet for the attackers, permanent), and then to affecting the system owners' ability to use military force to defend themselves. To dissuade attackers, the defender must then ask: what should be revealed to bolster the claim that its military missions have been hardened against all plausible effects of cyberwar?

A direct demonstration that defense systems are hard may not be plausible. Those who would demonstrate the quality of a battleship's armor, for instance, may show that a shell of a certain size did only a little damage. Potential adversaries may not have the same shells, but they could well extrapolate from their inventory to determine how much they could damage the battleship in question. But what cyber "weapons" would the defender have to make a demonstration against in order to demonstrate to others that their "weapons" would not work well? Detailing the exact nature of the weapons—the vulnerabilities such weapons were designed against and the exploits that worked against them—would only show that the military has been hardened against a known threat. It hardly answers the question of how well it works against an unknown — for example, zero-day—vulnerability, which, by its nature cannot be easily tested against. As a general rule, the cyber community takes as given that the ability to penetrate a system is the ability to bring it down. What is less obvious and has rarely been demonstrated is that the ability to take a system down implies the ability to keep a system down long enough to generate significant battlefield consequences.

DEMONSTRATING RESILIENCE

This then leads to the general argument in favor of being able to demonstrate resilience: that even a cyberattack with consequences cannot dent the military's capabilities. A military that can show that its people are sufficiently agile and its capabilities sufficiently robust against all manner of disasters also demonstrates that it enjoys similar immunity to disasters from cyberspace. That is, the point is not for the potential target to demonstrate its ability to block and/or neutralize an attack but to show that its missions can succeed even though the attack worked. Important elements of this argument would be the overall resilience of the system, the adaptability of warfighters, the diversity and redundancy of its communications paths, the error-tolerant features of its control and decision systems, the multiple layers of its error-detection and error-correction protocols, safety and fail-safe features of weapons systems, and the ability to revert to earlier less-sophisticated perhaps analog systems if digital ones stop working well.[18] In essence, what needs to be demonstrated is the quintessentially military nature of battlefield resiliency. At issue is not how well one side can defeat cyberattacks—for who can prove that the attacks were particularly strong? It is how well it can operate if cyberattacks do exactly what they were supposed to do— for who will quibble that the demonstration attacks were too strong? The ability to repel a series of created attacks perfectly, although laudable in practice, misses the point.

Many techniques of resilience are well-understood. One element is redundancy, the provisioning of stand-by capacity that is unused or little used normally but is available for emergencies. Another, correspondingly, is the ability to identify and suppress low-priority applications whose use of resources gets in the way of high-priority applications.[19] Diversity is another element in redundancy that comes in handy against attacks that target particular types of systems:[20] for example, satellites are used as stand-by for terrestrial telephony. A rapid response capability—such as the ability to deploy medical teams or evacuation teams—also helps particularly if the disaster threatens people's lives. Loose coupling[21] helps insulate complex systems against cascading failure through a combination of slack and circuit-breakers; this permits downstream systems to be separated from shocks in upstream systems (e.g., battery backup that permits an orderly shutdown of systems, such as aluminum smelters, when main power goes). Attitudes are an important component of resiliency. If power goes out, are people prepared and willing to move to tasks that do not require power? Do they own flashlights? Can they read maps? Testing is an oft-overlooked component of resiliency as well.[22] Many disasters take place because the backup capability deteriorated— and no one noticed until too late. Analysis and engineering matters; it provides an educated guess as to how much of what kind of resiliency to put into a complex system.

In sum, the best dissuasion against starting a war based on the adversary's hopes that operational cyberwar can be decisive is not demonstrating that operational cyberwar can be defeated on its own terms but rather demonstrating that even success at operational cyberwar is of little military consequence.

CONCLUSIONS

Denying the ability of hackers to attack systems is unlikely to dissuade attempts very much. The fundamental problem is one of conveying a system's defenses to attackers. Such defenses are not obvious to the attacker (even as the defender's vulnerabilities are often not obvious to the defender). To gauge such defenses requires probing the target system, but the costs involved in developing a successful probing capability are often most of the costs of attacking in the first place. To wit, having gone so far, why not complete the job? Psychological and institutional factors further diminish how much hackers may be discouraged by the prospects of failure. Demonstrated resilience, however, may substitute for good defenses in discouraging hackers. At the strategic level, a studied insouciance ought to discourage coercive attacks. At the operational level, denying the ability of hackers to change the correlation of military forces even if defense systems are compromised should go a long way to dissuading countries from attacking military systems and then launching off to war on the premise that the correlation of forces has been changed. In other words, whereas direct denial is difficult, indirect denial may be robust.

NOTES

1. Henry Kissinger argues, for instance, that Chinese military actions in Korea (1950–1953), India (1962), and Vietnam (1979) and demonstration-level military actions against Taiwan (1996) were undertaken primarily for psychological effect; see his *On China*.
2. See chapters 1, 5, 6, 7 in this volume.
3. The risk of punishment (as in deterrence by punishment) is the other aspect of such costs.
4. For an explanation by the head of the NSA's Tailored Access Operations of how nation-state hackers compromise networks, see "USENIX Enigma 2016 - NSA TAO Chief on Disrupting Nation State Hackers," https://www.youtube.com/watch?v=bDJb8WOJYdA.
5. As described in Gartzke and Lindsay, "Weaving Tangled Webs."
6. See chapters 5 and 7.
7. In essence, whatever *a priori* likelihood it assigned to the possibility that penetrating the target would require an infinite number of attempts, every failed attempt eliminates the possibility that penetrating the target could be penetrated after 1, 2, 3 ... N attempts, there raising the relative *a posteriori* likelihood that it would take an limitless number of attempts.
8. Here, the *a priori* logic is that there *is* a pathway in. Every failure eliminates one failure or maybe a larger class of failures, leaving it more likely that the next try will find the pathway, since the past failures are no longer in the search space.
9. A founder of CrowdStrike has argued that one Chinese cyberespionage group, after having tried and failed multiple times, found software from CrowdStrike on a system it had penetrated and forthwith ceased its efforts. Shalal, "U.S. firm CrowdStrike claims success."
10. See, for instance, Cormac Herley, "When Does Targeting Make Sense for an Attacker?"
11. See, for instance, Cheswick and Bellovin, *Firewalls and Internet Security*.
12. Brown, "Hacking of Tax Records."
13. "Its" may refer to the content-distribution-network that the ultimate target connects to the Internet through.
14. If the target system processes incoming messages (rather than just return a page of information) certain ill-formed messages may generate a dis-

proportionate amount of processing on the target's part unless the target has ways of filtering out such requests.

15. "Strategic" is used in the sense that the Air Force does which is to cause effects away from the battlefield. It is not used to mean "really important."

16. Schneier, "Refuse to be Terrorized." In fairness, his advice was meant to prevent what he felt were stupid and costly defensive policies vis-à-vis discouraging terrorists, many of who carry out their attacks not for the effect on the actions of the targets, but to gain status against their rivals.

17. Two caveats merit note. First, if the target has multiple adversaries with varied cyberwar capabilities and strategic objectives, it should remember that what impresses one may not impress the other. Second, the act communicating one's preparedness may also impel others to ask" "what is going on that they think they have to say this?"

18. As an example of a downwards adjustment (admittedly one taken before GPS has been attacked). See Prudente, "Seeing stars, again."

19. Some power contracts with consumers, for instance, let them dial back their air conditioning as electrical generators approach capacity.

20. See, for instance, Geer, et al., *CyberInsecurity*.

21. A theme of Perrow, *Normal Accidents*.

22. The key recommendation Defense Science Board March 2013 study, *Resilient Military Systems and the Advanced Cyber Threat*, was to ensure that some percentage of all key defense systems be provably disconnected from cyberspace so that they could survive the worst imaginable cyberattack. Although this was framed in terms of a minimal deterrence strategy, the study did not speak to its value in dissuading a first-hack-then-fight strategy.

CONCLUSION

INTO THE NEXT CENTURY

THE CHANGING THEORY AND PRACTICE OF DENIAL

Andreas Wenger and Alex Wilner

The role of denial in the theory and practice of deterrence has grown over time because of broader changes in the international system. Recognizing the increasing complexity of international security affairs towards the end of the twentieth century, policy and scholarly communities began to explore deterrence settings beyond interstate relations and the nuclear realm—settings that seemed much more conducive to the logic of denial. Now, in the twenty-first century, deterrence plays itself out in a multipolar system, in asymmetrical strategic interactions between state and non-state actors, and in a conflict environment that often lacks a clear threshold or geographic separation between low and high conflict. Deterrence is a highly context-specific endeavor. This concluding chapter provides a broad assessment and analysis of how deterrence by denial has evolved over time and how it has been shaped by the changing nature of conflict, crisis, and war, and by the context—nuclear, conventional, interstate, substate, and cyber—in which it has been applied. Ten broad themes are explored, including conceptual challenges to exploring the logic of denial; the conceptual linkages between denial, defense, and punishment; denial within dynamic and complex interactions; denial in traditional and nontraditional security engagements; denial in classical and extended deterrence theory; and denial from criminological, terrorism studies, and cybersecurity perspectives. Together, these themes suggest avenues for further research.

CONCEPTUAL CHALLENGES

There is little doubt among scholars and practitioners regarding the importance of denial in contemporary deterrence. Yet, to date, the theoretical logic of denial in highly different deterrence settings has remained underspecified. At issue is the need for a better understanding of the conceptual distinction and relationship between defense, deterrence by denial, and deterrence by punishment, all the more so because in practice, these concepts are usually applied in combination. Another key problem is the difficulty in assessing the way denial interacts with a broader set of coercive and non-coercive tools in an ever-wider peace-conflict loop and across tactical, operational, and strategic levels. Scholars need to develop more robust definitions and mechanisms before we can undertake more rigorous empirical analyses of denial tactics. Empirical analyses are still scarce and limited in scope due to a general lack of information and data on cases where challengers have aborted their plans and changed their behavior due to a perceived likelihood of failure.

The classic distinction between deterrence by punishment and deterrence by denial was introduced by Glenn Snyder at the height of the Cold War. Both mechanisms induce an adversary *not* to do something. However, each mechanism functions by a distinct logic: Deterrence by punishment manipulates the decision calculus of a challenger through the application of threats; deterrence by denial influences the adversary's behavior by reducing the perceived benefits they expect to gain.[1] Both mechanisms are attempts at manipulating an adversary's cost-benefit calculus, but they approach the issue in opposite ways: punishment adds cost, whereas denial takes away benefits. Neither of the two deterrence mechanisms can control outcomes, however, and this sets them apart from defense. With deterrence, whether the defender succeeds in manipulating their adversary's decision calculus is entirely dependent on the adversary. By contrast, Thomas Schelling's "pure defense" offers control over outcomes.[2] In a unilateral effort, force is prepared and used to defeat an attack. Thus, whereas deterrence is conceptualized as a bargaining tactic, defense focuses on the local problem at hand and gives little consideration to adversary response, as James Wirtz reminds us in his contribution to this volume.[3]

In punishment and denial, the deterrence mechanism unfolds in distinct patterns, a point made by many authors in this book. The punishment mechanism presumes that the adversary will make the first move. Deterrence by punishment works primarily by imposing costs on an adversary, and the effect of the deterrent is closely linked to the (perceived) severity of the punishment. Punishment offers ongoing coercion, and in a crisis situation threats of punishment maximize the manipulative leverage. However, one consequence is that crises can turn into contests of will in which the credibility of resolve emerges as a critical factor. In contrast to the punishment mechanism, the denial mechanism presumes that the defender will make the first move. Deterrence by denial works primarily by reducing the likelihood that an adversary will achieve their objectives; the deterrent effect is thus closely linked to the perceived likelihood of failure. With denial, cost calculations are far less important than they are in punishment. Denial offers a proportionate stance, and, especially in a general deterrence situation, maximizes the manipulative leverage of an *ex ante* demonstration of resolve. The flipside is that with denial, the critical factor is the challenger's judgment of how credible the defender's capability is.

Applied in isolation, denial does not function well. From a theoretical perspective, manipulative attempts at pure denial tend to come up short on three counts. First, denial allows a challenger to control cost. A challenger can keep probing, and over time this allows them to better understand the real strengths and weaknesses of the defender in any given local context. Second, denial invites a displacement of pressure. As a challenger realizes they will not gain benefits from a given context, they may be tempted to shift their attack vector to softer targets. Third, the risk calculus at the heart of deterrence by denial remains unclear. The more uncertain a defender is about their adversary's level of motivation and their perception of the quality of one's own defenses, the greater the risks in a stand-alone denial strategy. These three issues explain why denial is seldom applied alone. They also underscore the key value of denial in a strategy of conflict: Denial can act as a conceptual and practical complement to defense and punishment.

DENIAL AS A CONCEPTUAL AND PRACTICAL COMPLEMENT TO DEFENSE AND PUNISHMENT

The better one understands the distinctions between defense, punishment, and denial, the better our understanding will become of the limits of a pure denial strategy and of the key role of denial in a strategy that combines defensive and offensive elements. Denial is indirectly adversary-oriented and functions as a bridge between the self-centered logic of defense and the adversary-centered logic of punishment, as John Sawyer explores in his chapter.[4] Unlike defense, denial aims to manipulate the adversary's perception (although, as discussed earlier, it cannot fully control it). Unlike punishment, denial works primarily through probability rather than through cost. However, while theory allows us to discuss the three concepts as distinct from each other, in practice they are inseparable because ultimately their effects depend not on the defender's strategic intent but rather the adversary's perception and calculation.

Linking Defense and Denial

Denial and defense are closely related but conceptually distinct mechanisms. Defense is a self-centered activity aimed at protecting the defender's population, infrastructure, territory, and forces. Force is prepared and used to repel and defeat an attack, thereby putting a specific target beyond the adversary's reach.[5] The (local) effect of defense is relatively well understood and is contingent on the quality of the defender's knowledge of their adversary's usable forces. By contrast, denial is indirectly adversary-oriented, designed to manipulate the challenger's decision calculus. Denial uses the threat of defeat as a means of averting attacks before they occur. It can be applied to military and non-military goals. Denial tactics always attempt to induce the perception in the adversary's mind that a certain goal cannot be achieved, at least not at an acceptable cost. In contrast to defense, the effects of denial are difficult to measure, as they are contingent on the quality of the knowledge the defender has of the adversary's goals.

In practice, denial and defense complement each other. Defense tends to be the strategist's natural preference because it is the only mechanism that offers control. However, a defense strategy has major weaknesses, as every policymaker will immediately point out: First, it is costly and demands potentially huge *ex ante* commitments. Second, robust knowledge of the adversary's intent and capability is a precondition for defensive success. However, denial can partially compensate for the weaknesses of defense: Defenses are by nature imperfect and costly, and this makes attempts at denial an interesting option. Defenses need not to be perfect; they need only be good enough to convince a challenger that an attack will fail or be very costly.

However, one should not assess defense only in terms of its material effectiveness and should not assess denial only in terms of its perceptional effectiveness. Both defense and denial may have broader positive and/or negative side effects that can affect their overall value. Defense attempts and denial attempts may have no effect on the adversary, but they can have positive or negative side effects on domestic and/or alliance politics. Even ineffective defenses may provide reassurance to citizens and allies because they can highlight the political willingness of a defender to act. Conversely, highly effective defenses may come at too high a price if they interfere with commerce and trade or with civil liberties. Thus, defense and denial may have no effect on the defender, and they can have unintended effects on the challenger by raising the challenger's threat level and thereby feeding a security dilemma in unintended ways.

From a defender's perspective, the critical distinction between defense (i.e., blocking and resisting an adversary) and denial (i.e., purposefully inducing the adversary from proceeding) lies in the defender's original intentions. Yet the defender will never be able to conclusively establish what worked and what did not. To understand why an attack occurred or did not occur, one would have to determine how the challenger perceived and evaluated the quality of the defender's defenses and threats. Yet from a challenger's perspective, defense and denial are often inseparable. A pure defense strategy may have a deterrent effect on the challenger, whereas a denial strategy may have no effect at all. The critical factor here is whether a specific capability of the defender—real or imagined— is perceived by the adversary as a valid deterrent, regardless of the defender's intent. Therefore, in practical terms, for both sides denial and defense are inherently entwined because there will always be uncertainty about the motivation and rationality of the challenger, about their risk calculus, about their evaluation of the likelihood of denial, and about a set of broader domestic and international political factors that affect the relationship between defender and challenger.

To be effective, defense and denial tactics need to be combined convincingly. Denial tactics that are not anchored in sufficient defenses invite the challenger to probe for weaknesses and find the weakest link in the defense or to simply displace their attack to a set of softer targets. Also, because deterrence is an interactive bargaining strategy, the relation between defense and denial remains open to political manipulation. Any adversary will evaluate their prospects of an attack in relation to available alternative courses of action. If there is no better alternative, and if the challenger is highly motivated, even a high likelihood of failure might not deter them from going through with an attack. The fact that denial depends on the adversary's risk calculus at a time of their own choosing is one of its key weaknesses. This is why strategists often add a cost element by linking denial to punishment.

Linking Denial and Punishment

Denial and punishment work through different mechanisms, but they are often linked, both in theory and in practice. Denial deters the adversary from leveraging the potential benefits of an attack by manipulating their perception of the likelihood of failure should they go ahead with an attack. Thus, it is the defender who makes the first move by attempting to change the adversary's perception of the status quo. Punishment, by contrast, deters the adversary by increasing the cost of an action through the threat of retaliation. Thus, it is the challenger who makes the first move (which the defender has attempted to deter in advance through threats of punishment). Denial is an indirectly adversary-oriented strategy, and punishment is a fully adversary-centered strategy, while both strategies ultimately attempt to dissuade another party—albeit through different mechanisms—from undertaking an action that is perceived as detrimental to one's own interests.

Denial and punishment complement each other, and each poses key challenges that may or may not be overcome, depending on the specific situation. If they are appropriately combined, they may partially compensate for some of their individual weaknesses. There are two critical factors that determine the success of threats of punishment: First, punishment can only work if the attribution problem is solved; punishment strategies presuppose good intelligence and a solid understanding of the decision calculus of the challenger. Second, proportionality and discrimination in punishment become increasingly important as the deterrence setting moves from existential to less-than-existential threats. If the threatened retaliatory response is perceived by the wider community as grossly disproportionate or illegal, punishment may have significant audience cost, both at the domestic and at the international levels. Denial has the potential to compensate for these weaknesses to some extent. Depending on the specific characteristics of the deterrence setting, denial may offer a non-provocative and defensive demonstration of commitment and/or some protection against disparate threats.

At the same time, there are two critical challenges that determine the success of denial: First, denial invites an adversary to work around demonstrated defenses and to shift their pressure to softer targets. Moreover, there is no guarantee that all intended adversaries will properly pick up on a generic deterrence-by-denial message. Second, demonstrating defensive power is costly for the defender, and denial threats allow the challenger to assess what probabilities of failure might be acceptable and to thereby control their costs. Punishment can compensate for these weaknesses to some extent: Depending on any given deterrence setting, punishment can be applied to a large set of targets, and it gives the defender a cost function that the adversary cannot control.

Denial and punishment thus complement each other's weaknesses. Denial threats tend to be applied as general threats to a whole class of actors or actions; punishment threats tend to be targeted at a specific actor or action. Denial threats are communicated only broadly and become palpable through demonstration; punishment threats convey a tailored message to a specific target and are communicated via customized communication. Denial capabilities tend to protect a specific target set, and this opens up some room for the challenger to displace their targets; punishment capabilities can easily be retargeted by the defender, and this opens up some room for "reuse" against a potentially broad target set. Denial threats are potentially very expensive and socially and economically costly for the defender; punishment threats tend to be less costly due to their reusability, their offense-to-defense ratio, and their lower level of intrusiveness in the defender's own domestic social and economic affairs. When applied together effectively, these contrasts can mitigate the shortcomings evident in either approach.

DEFINING THE RIGHT MIX OF STRATEGY FOR EACH SPECIFIC DETERRENCE SETTING

In the complex and mostly less-than-existential deterrence settings of the twenty-first century, denial and punishment work well in tandem and are, therefore, often combined in practical terms. In the context of extended regional deterrence, and in the case of managing low conflict, deterrence by denial can act as a general safety net, whereas deterrence by punishment adds coercive leverage. The appropriate mix of denial and punishment will depend on the characteristics of a specific deterrence context. As a general rule, one can expect that the weaker the threat, the more asymmetric and diffuse the involved actors, and the more general the deterrence setting, the greater the importance of the denial components will become. If the key aim is to maintain the status quo in the face of ongoing low-level tension or conflict, denial tactics may offer specific advantages: They are non-provocative and cooperation-inducing in terms of multi-actor deterrence; and they are reassuring both in terms of domestic and alliance politics.

Conversely, the more imminent the threat, the more symmetric and clearly defined the involved actors, and the more specific the deterrence setting, the greater the importance of punishment components will become. If the aim is to (re)establish red lines of acceptable behavior, punishment components may offer specific advantages: They link unacceptable behavior to clear costs; and they may act as a learning mechanism through which the batteries of generic deterrence can be recharged, a point that is discussed by Dima Adamsky and Alex Wilner in their chapters.[6]

A BROAD APPLICATION OF DENIAL: COMPLEX CONTEXTS AND DYNAMIC INTERACTIONS

The obvious differences between traditional and nontraditional threats demand a broader and more context-specific application of denial. The scope conditions of traditional deterrence theory are too narrow to provide an appropriate conceptual framework for today's highly complex and dynamic deterrence settings. As a consequence, the theoretical concepts of deterrence and denial have been broadened along three axes: beyond high conflict, military instruments, and grand strategy. At the same time, at the practical level, deterrence and denial have become far more dynamic and targeted. This simultaneous broadening and diversification of deterrence comes with additional challenges because it confuses the theoretical meaning of deterrent behavior, complicates the manner in which deterrence is put into action, and compounds the methodological challenges of rigorous empirical analysis of deterrence and denial processes.

THE PEACE-CONFLICT LOOP: EXTENDED DETERRENCE AS REGIONAL SECURITY MANAGEMENT

Traditional deterrence theory during the Cold War concentrated on high conflict and conceptualized deterrence as an outcome-oriented process of war avoidance. Policymakers and strategists focused on how the status quo could be upheld in times of acute crisis through the application of threats of massive punishment. Within the extended deterrence framework of the Cold War alliances, the essential prerequisites of deterrence—communication, capability, commitment, credibility—were identified and calibrated first and foremost in the military realm of nuclear and conventional forces. However, since the end of the Cold War, the nature of extended deterrence has fundamentally changed. US and Western extended deterrence has been reorganized into a broader influence-oriented activity which can occur during times of peace, crisis, and conflict, and which is implemented with a broad set of information, economic, diplomatic, and military tools, as Jonathan Trexel shows in his chapter in this volume.[7] Extended deterrence has thus changed from a narrow tool of escalation control and war avoidance into a more comprehensive instrument of regional security management.

Extended deterrence has shifted from a instrument for maintaining national sovereignty, the territorial integrity of an alliance, and the strategic stability between the two superpowers to a much broader tool for shaping regional security architecture and supporting the building of regional orders. In this shift, the addition of a denial component to the strategic mix has played a key role. Thus, in response to the changing dynamics of the extended deterrence regime in Asia, the United States has increased the role of effective theater ballistic missile defense— emphasizing a denial component—in a combined approach together with nuclear weapons and conventional power-projection capabilities.[8]

In his chapter, Trexel discusses the strategic relationship between the DPRK and Japan within this larger regional deterrence architecture. His analysis zooms in on the role of Japan's multilayered missile defense system, which can be seen as a new capability for pure defense or as a key tool for a denial strategy, both at the bilateral level and at the regional level, thanks to the interoperability of the Japanese BMD system with the US BMD system. Trexel is interested in the bilateral relationship and in Japan's use of its BMD system for denial and in how this goes beyond deterring the outbreak of war. He convincingly shows how Japan's counter-coercion denial strategy has performed consistently according to the theorized mechanism of deterrence by denial. Japan's BMD capability has denied the DPRK the desired benefits of further coercive missile tests over Japanese territory; stabilized the strategic relationship between the two countries; and established some space for political cooperation. Trexel's analysis further shows that a robust empirical assessment of denial is possible, while he also recognizes the narrow scope—coercive behavior in peace time associated with general deterrence—and the unique setting—the exclusively defensive capabilities of Japan—of the specific case he discusses in his chapter.[9]

The Integration of the Tactical, Operational, and Strategic Levels: Tradeoffs and Side-effects

Traditional deterrence theory during the Cold War conceptualized deterrence as an overarching framework that could be used to study and influence the strategic relationship between the two superpowers. Strategic stability at the global level was insulated from low conflict and regional dynamics through the nuclear threshold and the geographic separation between regional and global theaters of war. Since the end of the Cold War, however, low conflict has become increasingly intermingled with strategic-level relationships between states because of the spread of global terrorism and stateless cyberattacks. In order to deal with such nontraditional asymmetric conflicts between states and non-state actors, deterrence has become one tool in ongoing counterterrorism campaigns and in attempts to better protect the socio-technical infrastructures of state, private, and civilian actors against cyberattack. And whereas deterrence has played a significant role in the response of Western strategy and policy to the new asymmetric threats, the logic of denial has played a far greater role than the logic of punishment.

Several scholars have attempted to demonstrate the usefulness of deterrence in combatting terrorism. In particular, James Smith and Brent Talbot, in their influential article, have considerably broadened the conceptual scope of denial in asymmetric settings.[10] According to their framework, in counterterrorism, denial can be applied beyond the tactical level, where it is most commonly associated with the denial of opportunity through the communicative effects of target-hardening efforts. Denial at the tactical level can be combined with capability denial at the operational level (by limiting access to resources such as recruits, weapons, and sanctuary) and with a denial of objectives at the strategic level (by delegitimizing an adversary's message). Applying this framework to the specific challenge of deterring nuclear terrorism, Smith, writing alone, has convincingly argued that the net effect of deterrence by denial built bottom-up may also be an important part of combating WMD terrorism.[11] As John Sawyer notes in his contribution to this book, however, denial at the operational and strategic levels is only loosely conceptualized. While the overall outcome of denial-based deterrence against WMD terrorism can be discussed in terms of the logic of denial, logical ambiguities in terms of the outlined mechanism at the operational and strategic level persist.[12] It is even more difficult, as we have argued elsewhere, to understand the interactions and tradeoffs between denial attempts against WMD terrorism and ongoing parallel counterterrorism campaigns against conventional terrorism, which lean heavily on the threat and/or active use of punishment.[13]

Conclusively proving the cumulative influence of military and non-military denial tactics across the tactical, operational, and strategic levels will remain challenging at best and probably elusive in practice. For this reason, most empirical analysis has focused on denial attempts at the tactical level against conventional terrorism, rather than against CBRN terrorism. From a theoretical point of view, one of the reasons for this is that the risk calculus of deterrence by denial is indeterminate. The relationship between the risk propensity of any given actor and the point at which the risk of failure becomes unacceptable remains underspecified. This is especially problematic in asymmetric settings and threat environments populated by diffuse groups that are often ill-defined and not well known. Such actors often use the communicative leverage of violent action to tempt state actors into overreacting, with the aim of consolidating their group and garnering community support for their political aims.[14] Highly motivated attackers with various cultural or ideological rationalities may accept a very low probability of success —even failure may be perceived as bringing some benefit to their cause.

The most robust work that has assessed the effectiveness of denial in counterterrorism has focused on how the reconfiguration of physical space and social context has changed would-be offenders' perception dividing.

of the risk of failure. Sawyer focuses on dividing walls as one of the simpler examples of dissuasion by denial. He shows how two cases—Northern Ireland and Israel—provide some evidence of the dissuasive effects of dividing walls; he also highlights how tricky it is, from a methodological and practical point of view, to control the impact of both sides' simultaneous efforts to disrupt the leadership of their opponent and to control wider environmental changes.[15] Stein and Levi discuss the theoretical insight of studies in criminology and of experimental and quasi-experimental work on the effect of visible target-hardening. Their findings are consistent with other studies that conceive of terrorism as a subset of criminal activity and as usually consisting of acts perpetrated by individuals. Scholars of terrorism, by contrast, generally conceptualize terrorism as an organizational act aimed at a political purpose.

Denial in Traditional Security Engagements

The growing interest in conventional deterrence and denial tactics by Western policymakers and scholars in the post–Cold War period reflects two mutually reinforcing trends: First, new military technologies for long-range coercion have spread rapidly over the past decades, with major consequences for deterrence dynamics among great powers, allies, and regional newcomers. Ballistic missile technologies have offered emerging regional powers an affordable tool for a wide range of coercive behavior below the threshold of nuclear retaliation. Conventional long-distance precision-strike capabilities have provided the United States with an option to expand the role of conventional (as opposed to nuclear) tools in the protection of allies and friends. Conversely, the development of anti-access/area-denial strike capabilities has allowed China and Russia to threaten a *fait accompli* strategy, a theme explored by Wirtz in his contribution, as a means of rapidly achieving limited military gains and then switching to a defensive posture.

Second, the credibility of US nuclear security guarantees among key allies in the Asia Pacific region and the Middle East has been eroded over the past decades due to broader changes in the global security environment. Since the end of the Cold War, attitudes of most Western societies towards nuclear weapons have changed considerably, as the nuclear abolishment movement has become politically more salient. In 2009, US president Barak Obama reintroduced disarmament into international politics with his Global Zero initiative, which resonated widely around the world.[16] More specifically, the US withdrawal of tactical nuclear weapons from South Korea caused considerable doubt among allies about the credibility of US commitment to the allies' defense. Together, these two trends have reshaped the role of Western deterrence strategies in important and sustained ways.[17]

The focus of the current debate is on the role of denial in extended deterrence settings at the regional level. As US and Western deterrence postures have changed, the insights of conventional deterrence theory—which had taken a back seat to nuclear deterrence theory during the Cold War—gained renewed attention. In conventional deterrence during the Cold War, the threat of a preemptive or retaliatory counterstrike had been conceptualized as a denial approach. Yet the dilemmas of such an approach have become far more pronounced in today's security environment, and this has triggered significant adjustments to the toolbox of extended deterrence as an instrument of regional security management. A critical component of this adaptation process has been the integration of denial, through active defenses such as missile defenses, into a combined denial strategy.

RENEWED INTEREST IN CLASSICAL CONVENTIONAL DETERRENCE THEORY

In the Cold War context of absolute nuclear deterrence, the need to use force was seen as a symptom of deterrence failure, as Dima Adamsky notes in his chapter.[18] Classical Western conventional deterrence thinking followed the same line of thought, because conventional deterrence thinking was closely linked to necessities of nuclear deterrence. It conceptualized the use of force as part of a bargaining framework aimed at de-escalation and early war termination below the nuclear threshold.[19] The key aim of conventional deterrence during the Cold War for the United States was to establish a position of escalation dominance, thus in effect maximizing the credibility of the alliance's nuclear threats. Historically, the debate about conventional deterrence intensified in periods when the credibility of nuclear threats as a means of deterring a non-nuclear attack deteriorated at the political level. For instance, concerns that the Soviets did not believe the US would defend West Berlin with nuclear weapons lay at the heart of the shift in US strategy from massive retaliation to flexible response.[20] In the 1980s, growing public doubt about the efficacy of nuclear deterrence in Europe triggered a renewed debate about how best to enhance conventional deterrence and thus compensate for the declining effectiveness of nuclear deterrence.[21]

The Limits of Pure Denial Strategies

The Cold War debate about conventional deterrence expounded the problems associated with pure denial strategies and outlined the advantages of combined strategies that integrate elements of denial and punishment. Most authors agree that conventional deterrence has usually meant deterrence by denial.[22] The starting point of conventional deterrence theory is the timeless observation that war is particularly likely if *blitzkrieg*-style victories are possible. The argument hinges on the insight that the local military balance is critical to the calculations of the challenger. If the defender is in a position to threaten a war of attrition, they will be able to deny the adversary the option of a quick victory. In such settings, the credibility question focuses on locally available conventional capabilities, rather than on the issue of resolve, as is the case in the nuclear setting, discussed earlier.

A force designed for denial is more able to engage in conventional conflict; if threats need to be implemented, such a force offers control, rather than the continuing coercion of punishment.[23] Yet at the same time, such a force makes it relatively easy for the adversary to calculate and control the costs of an attack, and this may undermine the credibility of a denial strategy. Therefore, a reliance on defensive means only is a weaker deterrent strategy than a strategy that combines a denial component with a retaliatory threat. Adding a conventional retaliatory option increases the force of the deterrent because it increases uncertainty in the adversary's calculations and ensures that they cannot control the costs in advance.[24] However, the obvious disadvantage of a combined strategy is that it increases the escalatory risks and may facilitate a process that leads from deterrence to the use of force and open conflict.

Combined Conventional Denial Embedded in Nuclear Deterrence

Anchored within the top-down logic of extended nuclear deterrence, Western conventional deterrence during the Cold War was designed to establish a position of escalation dominance, linking the defensive and retaliatory denial option with the logic of nuclear preemption. Limited conventional retaliation was designed to convince the other side that war would be unprofitable and should be terminated before it escalated to the nuclear level. The same logic applied to the potential demonstrative use of limited nuclear force, where initial strikes were designed to deter further escalation. In both situations, the initial (conventional and nuclear) strikes represented, in effect, threats of further strikes and were designed to deter escalation. In such limited warfare, Glenn Snyder notes, "the nuclear weapons held in reserve by either side constitute a deterrent against the other side's expanding the intensity and destructiveness of its war efforts."[25]

THE GROWING ROLE OF DENIAL IN EXTENDED DETERRENCE

Since the end of the Cold War, policymakers and strategists have become considerably more interested in the role of denial in regional deterrence. During the 1990s, the US debate about extended deterrence was dominated by a perceived proliferation of nuclear capabilities in states such as North

Korea, Libya, and Iran. More recently, the spread of ballistic missile, cyber, and space capabilities to emerging regional powers has caused additional concerns regarding the credibility of US security guarantees to regional allies and partners. Ballistic missile technologies, in particular, are proliferating rapidly at the regional level. These offer regional actors tools that span great distances and are much cheaper than active interceptors. Ballistic missiles can be used in a variety of coercive ways short of war, facilitating coercive behavior below the threshold of nuclear retaliation. In addition to these new challenges by regional anti-status quo powers, the military modernization of China and Russia has added another set of problems to US deterrence policies. The development of anti-access/ area-denial capabilities has raised the specter of a so-called *fait accompli* strategy, as suggested earlier. Last but not least, the growing salience of the nuclear abolishment movement in Western societies, combined with mounting doubts about US nuclear guarantees among key allies in Asia and the Middle East, demand a reconfiguration of the regional deterrence architecture in Europe, Asia, and the Middle East.

As a reaction to these developments, the United States has reorganized its deterrence posture over the past decade. Washington now conceptualizes modern deterrence operations as a systematic effort to exercise decisive influence over adversaries' decision-calculus in peacetime, crisis, and war. Deterrence is no longer static but is a combination of a wide array of military tools (e.g., nuclear, BMD, space, cyber, reconnaissance) and non-military tools (e.g., diplomacy, sanctions, law enforcement). Extended deterrence is seen as a means for keeping threats geographically far away and for managing the global and regional security order.[26] To achieve these objectives, the United States has made considerable changes to the mix of nuclear, conventional, and missile defense capabilities underpinning its deterrence posture. Thus, while the role of nuclear weapons has declined markedly, the role of conventional power-projection capabilities and effective theater ballistic missile defenses in extended deterrence has been increased considerably.[27] Wirtz discusses this larger debate, criticizing the US deterrence strategy as a static war-prevention strategy that lacks credibility in the eyes of past and present adversaries. Echoing Samuel Huntington's call in the 1980s to add a counterstrike option to the mix of conventional deterrence, Wirtz conceptualizes denial as a hybrid strategy that combines the use of deterrence to prevent war with actual plans to fight and win a war, should deterrence fail.[28]

Redefining the Mix of Denial Tools: Active Defense versus Counterstrike

The current debate about conventional deterrence focuses on how the mix between defensive and offensive denial tools might be redefined. The starting point for the debate is the observation that in deterring aggression against less-than-vital interests, conventional threats will likely be the only politically feasible threats. The principles of proportionality and discrimination allow for threats with conventional weapons only. At the same time, however, the credibility of conventional threats will likely remain contested for at least two reasons: First, the relatively modest power of most conventional weapons and the long timespan needed to prepare their offensive use offers the adversary considerable room for (asymmetric) countermeasures. Second, in today's normative environment, the demonstration of resolve—in addition to the *ex ante* demonstration of capability—has become more important because some adversaries believe that US and Western democracies have become overly sensitive to casualties of combat and collateral civilian damage. As a result, conventional threats provide more room for interpretative flexibility and, consequently, deterrence communications need to be more context-specific and culturally tailored now than was the case during the Cold War.

To maximize the credibility of capability, conventional deterrence places a premium on forward-deployed and long-distance strike forces, thus combining a denial component with a retaliatory threat. Unlike during the Cold War, however, the escalatory risk of a combined strategy is no longer tightly coupled with the top-down logic of nuclear deterrence. The addition of a preemptive attack to the defensive deterrence-by-denial component might leverage the deterrent effect of ambiguity and uncertainty. Yet at the same time, such an expansion increases the risk of sliding from threats of punishment to prevention and open conflict. Moreover, shifting the focus from coercion to control can entail huge audience costs, as the strategy of military regime change in Iraq in 2003 illustrates. If a combined denial strategy evolves into a tool of regional ordering beyond the protection of allies and partner, it can trigger a security dilemma that in the long term may undermine strategic stability in various ways.

If one recognizes the dilemmas and trade-offs involved in a combined denial strategy, we can see more clearly the advantages of denial through active defenses (like missile defense) over denial through the threat of a counterstrike: First, active defenses provide visible protection to populations and a damage limitation option for allies. Although such defenses will likely be imperfect, even a limited capability may offer important political advantages in support of upholding resolve and credibility, both at the levels of domestic and of alliance politics. Second, active defenses increase the uncertainty for adversary leaders without the escalatory risk of a counterstrike option. Third, where less-than-vital interests are at stake, active defenses are more likely to accord with the global norms of proportionality and discrimination that govern the use of force.

Evidence from Asia Pacific

The growing role of denial through active defenses is shown in Trexel's assessment of the strategic interaction between the DPRK and Japan.[29] Trexel concludes that Japan's multilayered-denial BMD system has been successful in limiting the coercive behavior of the DPRK. The DPRK, provocatively brandishing its nuclear-capable missiles over Japanese territory, threatened Japan below the nuclear threshold of US security guarantees. The subsequent withdrawal of US tactical nuclear weapons from South Korea, which delinked the physical manifestation of nuclear threats from conventional deterrence, further increased the DPRK's room for coercive maneuvers. During this sub-conflict phase of the bilateral relationship, the advantages of denial through active defenses were confirmed. Japan's BMD system was successful in making the DPRK's ballistic missile threat less effective; it forced the DPRK to redirect the flight paths of missile tests away from Japanese territory; and because it was non-provocative and purely defensive, it also proved helpful at the political level, allowing Japan to demonstrate its commitment to defense to its own population, while inducing the other side to move towards a political process.

The greater emphasis on BMD integration by the United States and Japan was one element of a broader reassurance campaign in the changing US extended deterrence framework. The United States move from an on-the-ground to an off-shore extended deterrence posture in South Korea increased the coercive space for DPRK threats below the threshold of long-range nuclear retaliation. In addition, the US withdrawal of its tactical nuclear weapons from South Korea led to political disappointments beyond the nation directly affected. This was evident in the increasingly problematic security relationship between the United States and Japan. Consequently, Japan began to move towards a more autonomous position, as some domestic actors voiced concerns that relying solely on active defenses might not be enough. As a result, to reassure allies and deter the DPRK, the United States leveraged conventional strike forces and effective theater ballistic defenses and it even signaled the possibility of nuclear sharing.

Redefining Strategic Stability

Finding an appropriate balance between active defenses and long-range conventional counterstrike capabilities not only has wide repercussions for extended deterrence regimes at the regional level—it is also critical for strategic stability among the great powers. Both China and Russia are worried that the new US triad of conventional strike capability, missile defense capability, and nuclear capability will undermine their own nuclear deterrent.[30] Conversely, the United States is concerned about the destabilizing effects that China's and Russia's anti-access/area-denial capabilities could have on US regional deterrent regimes. Both Russia and China prefer a strategic stability that remains anchored in mutual nuclear vulnerability. In contrast, the United States argues that in the twenty-first century, the requirements for strategic stability need to be rethought. Clearly, the way in which strategic stability will be organized at the bilateral and trilateral great-power level will have major repercussions for how extended deterrence regimes are established at the regional level, and vice versa.[31]

DENIAL IN NONTRADITIONAL SECURITY ENGAGEMENTS

In nontraditional asymmetric interactions between state and non-state actors, the tenets of deterrence are fundamentally different from those in more traditional deterrence settings in terms of their conceptual nature; they are also considerably more limited in terms of the scope of their practical applicability than those in traditional deterrence settings. On the one hand, they are fundamentally different in their conceptual nature because they reflect a criminological concept of deterrence as much as they reflect an IR concept of deterrence. On the other, they are quite limited in terms of the scope of their practical applicability because in cases of counterterrorism, counter-insurgency, and cybersecurity, deterrence tools are conceptualized as only one pillar of a broader strategy, alongside other coercive and non-coercive pillars of influence.

In ongoing low-level asymmetric conflict, there is considerably less room for punishment strategies and a more pronounced emphasis on denial strategies than in more traditional deterrence settings. Punishment strategies rest on the premise that challengers are known; however, in a counterterrorism or cybersecurity context, attribution is often difficult, if not impossible. Moreover, threats of punishment in low-conflict situations may often not be perceived as credible by a challenger or may even be counter-productive, for example, if broader audiences perceive them as disproportionate.[32] In low conflict that plays itself out in a civilian space, protection becomes central for the population, for key infrastructures, and for security services. This explains why active and passive defense have gained currency in recent strategic thought: Because defense alone is costly and imperfect, recent attempts to embed defensive measures into an indirectly adversary-oriented denial tactic are in line with conventional deterrence thinking, which is based on a combined approach.

In nontraditional security engagements beyond the state, the logic of denial seems to offer some intuitive advantages. First, denial tactics tend to be less escalatory and are therefore more easily integrated into a broader set of diplomatic and civilian influence tools. Second, denial tactics can more readily be conceptualized as a holistic crisis management process and expanded into prevention and post-attack mitigation. Loose security communities and public-private partnerships are more easily built around the logic of denial. Finally, denial tactics are more easily applied in a broader normative sense in that they are aimed at minimizing the social rewards of attempts by non-state actors to leverage low-level political violence. Yet tweaking a concept that originated in the high conflict of the Cold War so that it fits the realities of today's low conflict setting has not been possible without adding new conceptual tools that so far had been applied primarily to individuals rather than to groups.

The Growing Relevance of Criminological Conceptions of Deterrence

In situations where there are multiple and diffuse threats to multiple and diffuse targets, deterrence takes on some of the characteristics of crime prevention. It is therefore not surprising that the criminological concept of deterrence has become increasingly relevant to our understanding of deterrence processes in nontraditional security engagements. In nuclear deterrence, deterrence is *absolute* and *specific*; in the deterrence of crime and political violence, deterrence is *restrictive* and *general*, as Thomas Rid has illustrated.[33] In the nuclear setting of the Cold War, deterrence was conceptualized as absolute—because one single instance of deterrence failure was unacceptable—and as specific—because there was one primary recipient of deterrence, and this allowed for a relatively clear message about the nature of any likely retaliation.

In nontraditional asymmetric settings, it is unlikely that low-level violence will ever cease. This is also the case regarding crime, which is therefore conceptualized as restrictive rather than absolute; crime prevention is thus aimed at minimizing the quantity and quality of offenses in a specific jurisdiction. Given the problem of attribution, criminological deterrence is also general rather than specific. Punitive action in the past creates precedents for future crimes, causing potential offenders to reevaluate their intentions and thus reducing the rate of delinquency. To deter crime, the larger community must perceive any punitive action as proportional to the crime. Further, in crime prevention, the "certainty of apprehension" is a more effective deterrent than "the severity of the ensuing legal consequence."[34] In short, deterrence in crime prevention is conceptualized as a learning experience for the community and potential offenders through which a range of acceptable norms is established. In such a deterrence framework, the relative weight of offense and punishment in relation to defense and denial is reversed in contrast to traditional state-to-state deterrence settings. Punishment must be limited and geared towards the (re)establishment of normative rules in justiciable cases, whereas denial is the critical component in maintaining the normative rules and dissuading attempts by potential offenders to erode these.

Linking Denial to Risk and Resilience

The criminological conception of deterrence has been applied to terrorism studies in a variety of ways, as discussed in the next section. Yet in the context of low conflict, where often very little is known about the risk-calculus and decision-calculus of potentially very heterogeneous offenders, mechanisms that aim at manipulating an adversary's cost-benefit calculation seem not very promising tools of coercion. The attribution problem in cyberspace is even more pronounced than in terrorism

—which is politically motivated and in large parts still nationally oriented —because the motivation of a cyber-perpetrator may be recreational or financial, rather than political, and because cyberspace is a globally interconnected space. This explains why in counterterrorism and cyber-security, the logic of denial has been combined with elements from the study of risk and resilience. Risk and resilience are multifaceted concepts and had been applied widely before they were integrated into social science research, in general, and into the field of disaster risk reduction, in particular.[35] Both risk and resilience, like defense, concep-tualize cybersecurity and counterterrorism as a self-centered activity. Unlike defense, however, they encompass the whole risk management process, from prevention to intervention to reconstruction (risk), and they conceptualize security and safety as a shared responsibility with shared costs that should be borne by the public, private, and civil sectors (resilience). Against this background, Patrick Morgan, in his contribution, rightly questions the wisdom of applying denial comprehensively to the whole panoply of so-called new threats. Following Morgan's insight, the following section discusses the many faces of denial in counterterrorism as distinct from denial in cybersecurity and cyber defense.

THE MANY SIDES OF DENIAL IN COUNTERTERRORISM

The application of denial theory to terrorism and counterterrorism has produced a rich palette of research. Yet the scope condition of the literature is characterized by high variance and considerable heterogeneity in terms of key concepts, research methods, and deterrence contexts. Some of the most robust work has concentrated on the simplest form of denial in counterterrorism: The reconfiguration of physical space, often labeled as target hardening, as reviewed by both Sawyer and Stein and Levi. Both chapters make a convincing argument that tactical denial through target hardening may contribute to a decrease in the probability of attack.[36]

Particularly interesting are the ways in which insights from criminology have been applied to terrorism and counterterrorism. Efforts to combat terrorism are in some ways comparable to efforts to fight crime: Both fields accept that the intent of perpetrators reflects the idiosyncrasies of a specific context, and both conceptualize their interventions as interactive activities that target multiple audiences. But at the same time there are major differences between fighting crime and combating terrorism: Crime prevention tends to focus on the individual criminal and acts within a given domestic jurisdiction, while large parts of counterterrorism are applied outside of judicial mechanisms. It makes sense, therefore, to distinguish between studies that analyze terrorism through a crime model and those that analyze terrorism through a war model.

Denial in a Domestic Crime and Terrorism Context

Denial-based approaches may be especially helpful in the context of home-grown and ethno-nationalist groups using terrorism, and less so for globally networked groups inspired by a transnational ideology.[37] In democracies, criminal punishment must follow the criminal-justice model. De-radicalization programs target local support networks and social contexts and contribute to the upholding of local norms. The denial of opportunity at critical targets nicely fits this domestic context. But at least two critical questions linked to denial efforts in such a context remain open for debate and further study: First, what kind of target hardening works best to counteract local terrorism? Second, does situational crime/terrorism prevention not simply force adversaries to displace an activity to another, softer target?

Analyzing terrorism through the framework of criminal activity allows Janice Gross Stein and Ron Levi to assess specific counterterrorism policies by using available evidence from crime prevention. Contrary to popular belief, crime displacement research shows that crime is less uniform and fungible than has often been assumed. Targeted police efforts at implementing local opportunity denial do not simply displace crime to another location.[38] Stein and Levi conclude that there is little empirical evidence in both crime and terrorism studies to suggest that would-be-offenders simply shift their activities to less well defended targets. On the contrary, within a domestic setting, denial approaches may have positive effects beyond securing a specific location.

Denial in a Transnational War on Terrorism Context

In his chapter on denial in Israeli thinking, Adamsky explains how a criminological conception of deterrence has informed Israeli strategy in dealing with a threat spectrum that, for a long time, has loosely linked external threats to internal ones, military campaigns to terrorist campaigns, and state adversaries to non-state adversaries. Israel has conceptualized its repeated victorious use of force on the battlefield as an integral component of deterrence and as necessary to recharge the batteries of deterrence and postpone the next round of violence. This traditional deterrence conception was adapted after the Second Intifada, as the threat transformed into an asymmetric mix of insurgency, suicide terrorism, and WMD proliferation. Because decisive battlefield decisions were no longer possible, an adapted Israeli paradigm emerged in which deterrence became the cumulative result of intensive short operations and ongoing limited operations. While it was unlikely that levels of violence would ever reach zero, precedents of punitive offensive action would help reduce the rate of attacks—a thought comparable to a criminological conception of deterrence as a learning experience designed to establish and sustain acceptable norms.

More recently, the Israeli strategic community has begun to appreciate the value of active and passive defenses and has integrated these into a combined punishment and denial strategy. Since in an asymmetric environment the proportional application of force is important, the combination of episodic punishment and denial has offered Israel domestic and international political advantages. The two critical denial components are the building of the Israeli security fence (passive defense) and, in reaction to a growing missile threat, the construction of a layered BMD system (active defense). The latter made a more restrictive application of force more acceptable, while simultaneously increasing a feeling of protection among the population. Paradoxically, Adamsky concludes, while Israeli attempts at denial might have had a strong effect on Israeli adversaries, Israel's strategic community remains skeptical about the value of defensive measures. Too strongly ingrained in Israeli strategic culture is the belief that technology favors the offense (ballistic missiles rather than ballistic missile defense) and denial leads to displacement (the fence may have impeded suicide bombings but ultimately led to missile and rocket attacks).[39]

DENIAL IN CYBERSECURITY AND CYBER DEFENSE

Denial in counterterrorism has many faces when one considers the variety of legal and political conflict contexts. However, by comparison, denial in cybersecurity and cyber defense spans an even wider spectrum of threats, including recreational hacking, cybercrime, cyber espionage, cyber terrorism, operational cyberattack, and strategic cyberwar.[40] Martin Libicki, in his contribution, focuses primarily on globe-spanning actors at the higher end of the conflict spectrum.[41] At the same time, the majority of cyber incidents still fall into the category of low end cybercrime. As a consequence of this very broad threat spectrum, a comprehensive cybersecurity policy must integrate a wide variety of tasks—cyber intelligence, cyber governance, cyber resilience, cyber law enforcements, cyber defense, international cybersecurity policy, IT-security, and digital awareness.[42] These are implemented across different legal settings by a broad set of public, private, and civilian actors. Within a comprehensive approach to cybersecurity and cyber defense, denial takes on a multidimensional role, reflecting at least three distinct tasks and legal settings.

Cyber Resilience and Critical Infrastructure Protection

For some time, there has been a trend towards resilience-based approaches to cybersecurity. This reflects a general focus on protecting critical infrastructures against lower level attacks. Resilience approaches, as a self-centered and shared activity of the public, private, and civil sectors, have seemed to be the most sensible way to harden infrastructure and make communities resilient.[43] Given the difficulty of attribution and the practical problems in conveying the strength of a system's defense to potential attackers, there is little room for targeted denial approaches in cyberspace, a point underlined by Libicki.[44] Techniques such as firewalls and antivirus software as means to secure communications networks, information networks, and infrastructures are passive. By contrast, however, resilience approaches are more anticipatory, as they concentrate on system continuity through backups, redundancies, awareness campaigns, and information sharing between public and private actors.[45] In this way, resilience approaches can adopt some elements of (indirect) denial, especially as regards their more strategic, preventative, and communicative elements. In other words, denial can contribute to cybersecurity through repeated demonstrations of failure, which signal the capacity for continued service.

Operational Cyberattacks in Support of Kinetic Attack

The cyber domain has been widely recognized as the fifth domain of warfare along land, sea, air, and space dimensions. Computer network attacks are now a key component of modern warfare. In response to the politicization and militarization of cyberspace over the past few years, active and offensive approaches have been combined, and priority has been given to identifying foreign perpetrators in advance, as part of a peace-time cyber defense policy. In his contribution, Libicki reminds us that cyberattacks are rarely an end in themselves because they cannot disarm a force or seize assets in the way other weapons can. Cyberattacks in military operations are often instrumental, and their aim is to create opportunities for kinetic attacks. The critical issue here is that cyberattacks may be destabilizing, potentially increasing both the attacker's and the defender's temptation to escalate.[46] Deterring cyberattacks through attempts at direct denial would therefore be desirable. Yet implementing direct denial is difficult, if not impossible, as Libicki points out, because an attacker can take down a system only if they are aware of a vulnerability that the defender is not aware of.[47] A better way to convince an attacker that one's military operations are not at risk from cyberattack is to demonstrate resilience. Protecting one's system from an attack is one thing; demonstrating that the system is resilient and that one's missions will succeed despite such an attack is more complex and relates to the logic of indirect denial through resilience.

Cyber Espionage Versus Denial through Disclosure

Libicki makes the point that in cyberwar "a vulnerability discovered is a vulnerability neutralized."[48] Targeted denial in such a context is difficult to implement because most of an attacker's investment is a generic investment in zero-day vulnerability, rather than an investment in reconnoitering a specific target. Can states turn this feature into an effective denial strategy, following the logic that a vulnerability disclosed is an attack option neutralized? Public disclosure of the technical details of other countries' cyberattacks would neutralize an attack option and increase the attacker's costs for developing new attack options. As Morgan points out, the purpose of many high-end cyberattacks is to gain intelligence, and this may mean that countries are reluctant to release information about zero-day vulnerabilities. Typically, large states that are among the most vulnerable to high-end cyberattacks are often also among the biggest perpetrators of cyberattacks through their extensive cyber espionage activities. States need to balance the benefits of keeping intelligence secret, on the one hand, and publicly disclosing it, on the other. Public disclosure may send signals to adversaries that affect their cost calculation. Recently, an increasing number of states have chosen to publish information about adversarial cyber operations discovered in their networks. Little is known about their motivation so far, but the trend of increased disclosure may indicate that some states are thinking about ways to strengthen their denial capacities against high-end attacks.

FINAL THOUGHTS: NEXT STEPS FOR RESEARCH

Several questions arise from this volume—and are left unanswered to highlight where research on deterrence might be heading in the future. First, as the boundaries of deterrence theory are stretched and thinned in innumerable ways to accommodate the rising prevalence of denial in nontraditional security environments, will the meaning of deterrence also take on new parameters? At what point does a continued broadening of deterrence in theory dilute the fundamental principles of deterrence in practice?[49] This volume has illustrated how and why the processes of defense, punishment, and denial are inherently linked, but how are influence, deterrence, and coercion linked, both conceptually and practically? What are the risks and promises of blending these concepts together?

Second, criminological deterrence, to a certain extent, naturalizes the occurrence of crime: The baseline is that crime is unwanted but also inescapable. Crime can be managed, tamed, and perhaps manipulated, but it can never be truly eliminated. Does borrowing the logic of criminological deterrence for use in counterterrorism, cybersecurity theory, and practice also necessitate the absorption of this baseline logic? Do we risk naturalizing terrorism and cyberattacks by treating their deterrence as but one shifting point on a spectrum? If so, what political and social effects might this have on the ways in which states and societies think about, contend with, and counter both terrorism and cyberattacks?

Third, deterrence has long been difficult to study empirically. Out of necessity, deterrence research often dabbles in non-events: the assault not mounted; the attack never carried out. How do the lessons of the empirical examination of denial stemming from this volume lend themselves to the broader study of deterrence? For instance, are denial and punishment, despite their inter-linkages, so unique in practice that empirical lessons from one might not translate into the other? What empirical pitfalls and promises do punishment and denial share? Fourth, as this volume illustrates, the scope of denial reaches well beyond the military and kinetic domains. In theory, denial, when woven together with punishment, uses diplomacy, statecraft, intelligence, economics, and other methods to shape and manipulate an adversary's or challenger's behavior. In practical terms, what are the limitations of shifting between the kinetic and non-kinetic domains of deterrence? If the military and civilian leaders of defense departments give priority to deterrence by punishment and retaliation, what equivalent government department, agency, or body directs and controls a denial strategy? And in knitting punishment together with denial, what conflict of interests might emerge between these disparate civilian and military leaders?

NOTES

1. Snyder, *Deterrence and Defense*, 4, 9–11, and 14–15.
2. Schelling, *Arms and Influence* (2nd ed.), 78–19.
3. See chapter 5.
4. See chapter 4.
5. Auerswald, "Deterring Nonstate WMD Attacks," 545–548.
6. See chapters 7 and 2.
7. See chapter 6; Morgan, "The State of Deterrence in International Politics Today."
8. von Hlatky and Wenger, *The Future of Extended Deterrence: The United States, NATO, and Beyond* (Washington, DC: Georgetown University Press, 2015).
9. See chapter 6.
10. Smith and Talbot, "Terrorism and Deterrence by Denial," 16–17.
11. Smith, "Strategic Analysis, WMD Terrorism."
12. See chapter 4.
13. Wenger and Wilner, "Deterring Terrorism: Moving Forward."
14. Spaniel, "Rational Overreaction to Terrorism."
15. See chapter 4.
16. Obama, "Remarks of President Barack Obama."
17. See chapter 6.
18. See chapter 7.
19. Gerson, "Conventional Deterrence in the Second Nuclear Age"; Stone, "Conventional Deterrence and the Challenge of Credibility."
20. Wenger, *Living with Peril*.
21. Huntington, "Conventional Deterrence and Conventional Retaliation."
22. Huntington, "Conventional Deterrence and Conventional Retaliation"; John Mearsheimer, "Why the Soviets Can't Win Quickly"; Gerson, "Conventional Deterrence in the Second Nuclear Age"; Stone, "Conventional Deterrence and the Challenge of Credibility."
23. Freedman, *Deterrence*, 39.
24. Huntington, "Conventional Deterrence and Conventional Retaliation."
25. Snyder, "Deterrence and Power."
26. Morgan, "State of Deterrence."
27. Wenger, "Conclusion: Reconciling Alliance Cohesion." .
28. See chapter 5.

29. See chapter 6.
30. Bernstein, "Ballistic Missile Defense in Europe"; Roberts, "Extended Deterrence and Strategic Stability."
31. Dunn, "Building Toward a Stable and Cooperative Long-Term U.S.–China Strategic Relationship"; Colby and Gerson, *Strategic Stability*.
32. Wenger and Wilner, *Deterring Terrorism*.
33. Rid, "Deterrence Beyond the State"; Tor, "'Cumulative Deterrence' as a New Paradigm."
34. Nagin, "Deterrence in the Twenty-First Century."
35. Fleming and Ledogar, "Resilience, an Evolving Concept"; David Alexander, "Resilience and disaster risk reduction."
36. See chapters 3 and 4.
37. Gearson, "Deterring Conventional Terrorism."
38. See chapter 3; Bowers and Johnson, "Measuring the Geographical Displacement."
39. See chapter 7.
40. Dunn Cavelty, "Cyberwar," 123–144.
41. See chapter 8.
42. Nye, "Deterrence and Dissuasion in Cyberspace."
43. Demchak, "Resilience and Cyberspace"; Stevens, "A Cyberwar of Ideas?"
44. See chapter 8.
45. Dewar, "Active Cyber Defense."
46. Rebecca Slayton, "What Is the Cyber Offense-Defense Balance?"
47. See chapter 8.
48. Ibid.
49. Wilner, "Contemporary Deterrence Theory and Counterterrorism."

BIBLIOGRAPHY

Abmann, Lars. *Theater Missile Defense (TMD) in East Asia: Implications for Beijing and Tokyo*. Berlin: LIT, 2007.

Abrahms, Max. "What Terrorists Really Want: Terrorist Motives and Counterterrorism Strategy." *International Security* 32, no. 4 (2008).

Acton, James. *Silver Bullet? Asking the Right Questions about Conventional Prompt Global Strike*. Washington, DC: The Carnegie Endowment, 2013.

Adamsky, Dima. *The Culture of Military Innovation*. Stanford: Stanford University Press, 2010.

Alexander, David. "Resilience and disaster risk reduction: an etymological journey." *Natural Hazards and Earth System Sciences* 13, no. 11 (2013).

Allen, Greg, and Taniel Chan. "Artificial Intelligence and National Security." Belfer Center Study. Harvard Kennedy School, 2017.

Almog, Doron. "Cumulative Deterrence and the War on Terrorism." *Parameters* 34, no. 4 (2004).

Amidror, Yaakov. "Missile Defense: An Israeli Perspective." INSS, January 2014.

Analysans. "North Korea's September 15 Hwasong-12 Test: The Why and The Wherefore." September 2017.

Anthony, Robert. *Deterrence and the 9-11 Terrorists*. Alexandria, VA: Institute for Defense Analysis, 2003.

Arce M., Daniel G., and Todd Sandler. "Counterterrorism: A Game-Theoretic Analysis." *Journal of Conflict Resolution* 49:2 (2005).

Art, Robert, and Pat Cronin, eds. *The United States and Coercive Diplomacy*. Washington, DC: US Institute of Peace, 2003.

Associated Press. "Japan deploys PAC-3 missile interceptor near North Korea flight path," September 19, 2017.

Atzili, Boaz, and Wendy Pearlman. "Triadic Deterrence: Coercing Strength, Beaten by Weakness." *Security Studies* 21 (2012): 301–335.

Auerswald, David. "Deterring Nonstate WMD Attacks." *Political Science Quarterly* 121, no. 4 (2006).

Baker, James, III. *The Politics of Diplomacy.* New York: GP Putnam's Sons, 1995.

Bapat, Navin. "State Bargaining with Transnational Terrorist Groups," *International Studies Quarterly* 50, no. 1 (2006).

Bar, Shmuel. "Deterrence of Palestinian Terrorism: The Israeli Experience." In *Deterring Terrorism: Theory and Practice*, edited by in Andreas Wenger and Alex Wilner. Stanford: Stanford University Press, 2012.

Bar-Joseph, Uri. "Variations on a Theme: The Conceptualization of Deterrence in Israeli Strategic Thinking." *Security Studies* 7, no. 3 (1998).

BBC News. "Three Guilty of Airline Bomb Plot." September 6, 2009.

Beckley, Michael. "The Emerging Military Balance in East Asia." *International Security* 42, no. 2 (2017).

Bernstein, Paul. "Ballistic Missile Defense in Europe: Getting to Yes with Moscow?" In *The Future of Extended Deterrence: The United States, NATO, and Beyond*, edited by Stéfanie von Hlatky and Andreas Wenger. Washington, DC: Georgetown University Press, 2015.

Betts, Richard. "The Soft Underbelly of American Primacy: Tactical Advantages of Terror." *Political Science Quarterly* 117, no.1 (2002).

Blakemore, Sarah-Jayne, and Jean Decety. "From the Perception of Action to the Understanding of Intention." *Nature Reviews Neuroscience* 2 (2001).

Blumstein, Alfred, Jacqueline Cohen, Jeffrey Roth, and Christy Visher, eds. *Criminal Careers and "Career Criminals."* Vol. 1. Washington, DC: National Academy Press, 1986.

Bluth, Christoph. *Korea.* Malden: Polity Press, 2008.

Bowers, Katie, and Shane Johnson. "Measuring the Geographical Displacement and Diffusion of Benefit Effects of Crime Prevention Activity." *Journal of Quantitative Criminology* 19, no. 3 (2003).

———, et al. *Systematic Review Protocol: Spatial Displacement and Diffusion of Benefits Among Geographically Focused Policing Initiatives.* Oslo: Campbell Systematic Reviews (2011).

Braga, Anthony. "Getting Deterrence Right? Evaluation Evidence and Complementary Crime Control Mechanisms." *Criminology and Public Policy* 11 (2012).

———. "Title Registration for A Review Proposal: Broken Windows Policing to Reduce Crime in Neighborhoods." Unpublished Manuscript.

———, Andrew Papchristos, and David Hureau. "The Effects of Hot Spots Policing on Crime: An Updated Systematic Review and Meta-Analysis." *Justice Quarterly* 31 (2012).

———, and David Weisburd. "The Effects Of "Pulling Levers" Focused Deterrence Strategies on Crime." Oslo: Campbell Systematic Reviews, 2012.

Bratton, William, and Peter Knobler. *The Turnaround: How America's Top Cop Reversed the Crime Epidemic.* New York: Random House, 1998.

Brewer, William, and Bruce Lambert, "The Theory-Ladenness of Observation and the Theory-Ladenness of the Rest of the Scientific Process." *Philosophy of Science* 68, no. 3 (2001).

Brimmer, Esther, and Daniel Hamilton. "Introduction: Five Dimensions of Homeland and International Security." In *Five Dimensions of Homeland and International Security*, edited by Esther Brimmer. Washington, DC: Center for Transatlantic Relations, 2008.

Brom, Shlomo. "Israel's Missile Defense: Pros and Cons." INSS, January 2014.

Brown, Robbie. "Hacking of Tax Records Has Put States on Guard." *New York Times*, November 5, 2012.

Brun, Itai. "The Other Revolution in Military Affairs." *Journal of Strategic Studies* 33, no. 4 (2010): 535–567.

Buchanan, Ben. *The Cybersecurity Dilemma: Hacking, Trust, and Fear Between Nations.* Oxford: Oxford University Press, 2017.

Bueno de Mesquita, Ethan. "The Quality of Terror." *American Journal of Political Science* 49, no. 3 (2005).

Bush, George H. W. "A Warning Letter to Saddam Hussein." In *Deterrence and Saddam Hussein: Lessons from the 1990–1991 Gulf War,* by Barry Schneider. The Counterproliferation Papers, Future Warfare Series No. 47. Maxwell AFB: USAF Counterproliferation Center, 2009.

Cantor, David, and Kenneth Land. "Unemployment and Crime Rates in the Post-World War II United States." *American Sociological Review* 50, no. 3 (1985).

Caplan, Joel, et al. "Police-Monitored CCTV Cameras in Newark, NJ." *Journal of Experimental Criminology* 7, no. 3 (2011).

CBS News. "North Korea missile tests – a timeline." September 6, 2017.

Center for Strategic and International Studies. "North Korean Missile Launches & Nuclear Tests: 1984–Present," November 2017.

Cha, Victor. *The Impossible State: North Korea, Past and Future.* New York: HarperCollins Publishers, 2012.

Chalmers, Malcolm. "Deterrence and Counterproliferation." In *Deterrence in the twenty-first century: proceedings, London, UK, 18-19 May 2009.* Maxwell AFB: Air University Press, 2010.

Chamberlain, Dianne Pfundstein. *Cheap Threats: Why the United States Struggles to Coerce Weak States.* Georgetown: Georgetown University Press, 2016.

Cheswick, William, and Steven Bellovin. *Firewalls and Internet Security: Repelling the Wily Hacker.* New York: Addison-Wesley, 1994.

Chilton, Kevin, and Greg Weaver. "Waging Deterrence in the Twenty-first Century." *Strategic Studies Quarterly* 3, no. 1 (2009).

Clarke, Ronald, and Graeme Newman. *Outsmarting the Terrorists.* New York: Praeger, 2006.

Colby, Elbridge, and Michael Gerson, eds. *Strategic Stability: Contending Interpretations.* Carlisle, PA: U.S. Army War College Press, 2013.

Cook, Phillip. "The Impact of Drug Market Pulling Levers Policing on Neighborhood Violence: An Evaluation of the High Point Drug Market Intervention." *Criminology and Public Policy* 11, no. 2 (2012).

Corman, Hope, and Naci Mocan. "Carrots, Sticks, And Broken Windows." National Bureau of Economic Research, Working Paper No. 9061 (2002).

Cragin, Kim, and Scott Gerwehr. *Dissuading Terror: Strategic Influence and the Struggle Against Terrorism.* Washington, DC: RAND, 2005.

Crelinsten, Ronald. "Perspectives on Counterterrorism: From Stovepipes to a Comprehensive Approach." *Perspectives on Terrorism* 8, no. 1 (2014).

Crenshaw, Martha. "Will Threats Deter Nuclear Terrorism?" In *Deterring Terrorism: Theory and Practice*, edited by Andreas Wenger and Alex Wilner. Stanford: Stanford University Press, 2012.

Cronin, Audrey Kurth. *How Terrorism Ends: Understanding the Decline and Demise of Terrorist Campaigns.* Princeton: Princeton University Press, 2009.

Davis, Paul. "Deterring and Otherwise Influencing Violent Extremist Organizations (VEOs)." Presentation at Office of the Secretary of Defense Strategic Multilayer Assessment meeting. Bethesda, MD, 2011.

———. "Towards an Analytical Basis for Influence Strategy in Counterterrorism." In *Deterring Terrorism: Theory and Practice*, edited by Andreas Wenger and Alex Wilner. Stanford: Stanford University Press, 2012.

———, and Brian Jenkins. *Deterrence and Influence in Counterterrorism.* Washington DC: RAND, 2002.

———, and John Arquilla. *Deterring or Coercing Opponents in Crisis: Lessons from the War with Saddam Hussein.* Santa Monica: RAND, 1991.

Dawson, Chester. "Japan Shows Off Its Missile-Defense System." *Wall Street Journal* December 9, 2012.

Defense Intelligence Agency. "Ballistic and Cruise Missile Threat 2017," 2017.

Defense Science Board March. *Resilient Military Systems and the Advanced Cyber Threat.* 2003.

Demchak, Chris. "Resilience and Cyberspace: Recognizing the Challenges of a Global Socio-Cyber Infrastructure." *Journal of Comparative Policy Analysis: Research and Practice* 14, no. 3 (2012).

Deni, John. "Modifying America's Forward Presence in Eastern Europe." *Parameters* 46, no. 1 (2016).

Department of Defense. *Ballistic Missile Defense Review Report.* Washington, DC: Department of Defense, 2009.

———. *Missile Defense Review.* Washington, DC: Department of Defense, 2019.

———. *Quadrennial Roles and Missions Review Report.* Washington, DC: Department of Defense, 2009.

Department of Homeland Security. *National Preparedness Goal.* Washington DC: D of Homeland Security, 2011.

Department of State. *Fact Sheet: Missile Defense and Deterrence.* Washington, DC: Department of State, 2001.

Dewar, Robert. "Active Cyber Defense." *CSS Cyber Defence Trend Analysis.* Zurich: Center for Security Studies, 2017.

Dombrowski, Peter. "Demystifying the Iron Dome." *The National Interest* no.126 (July 2013): 49–59.

Dunn Cavelty, Myriam. "Cyberwar." In *The Ashgate Research Companion to Modern Warfare*, edited by George Kassimeris and George Buckley. Farnham: Ashgate, 2010.

Dunn, Lewis, ed. "Building Toward a Stable and Cooperative Long-Term U.S.–China Strategic Relationship: Results of a Track 2 Joint Study by U.S. and Chinese Experts." *Issues & Insights* 13, no. 2 (2012).

Dutter, Lee, and Ofira Seliktar. "To Martyr or Not to Martyr: Jihad Is the Question, What Policy Is the Answer?" *Studies in Conflict & Terrorism* 30, no. 5 (2007).

Eck, John. "The Threat of Crime Displacement." *Criminal Justice Abstracts* 25 (1993).

———, and Edward Maguire. "Have Changes in Policing Reduced Violent Crime?" In *The Crime Drop in America*, edited by Alfred Blumstein and Joel Wallman. Cambridge: Cambridge University Press, 2000.

Eligon, John. "Taking on Police Tactic, Critics Hit Racial Divide." *New York Times*, March 22, 2012.

Elleman, Michael. "North Korea's Hwasong-12 Launch: A Disturbing Development." *38 North*, August 30, 2017.

Enders, Walter, and Todd Sandler. *The Political Economy of Terrorism.* Cambridge: Cambridge University Press, 2011.

———, Todd Sandler, and Jon Cauley. "Assessing the Impact of Terrorist-Thwarting Policies: An Intervention Time Series Approach." *Defence and Peace Economics* 2, no. 1 (1990).

Eran, Ehud. "Israel and Weak Neighboring States." *Mitvim* (February 2013): 6–8.

Erickson, Maynard, and Jack Gibbs. "Specific Versus General Properties of Legal Punishments and Deterrence." *Social Science Quarterly* 56 (1975).

Farrell, Graham, Sylvia Chenery, and Ken Pease. "Consolidating Police Crackdowns: Findings from an Anti-Burglary Project." Police Research Series Paper No. 113 (1998).

Farrington, David, et., al., "The Effects of Closed-Circuit Television on Crime" *Journal of Experimental Criminology* 3, no. 1 (2007).

Feldman, Shai. "Deterrence and the Israeli-Hezbollah War" In *Deterrence in the 21st Century*, edited by Anthony Cain. Maxwell AFB: Muir S. Fairchild Research Information Center, 2009.

Fleming, John, and Robert Ledogar. "Resilience, an Evolving Concept: A Review of Literature Relevant to Aboriginal Research." *Pimatisiwin* 6, no. 2 (2008).

Freedman, Lawrence. *Deterrence.* London: Polity, 2004.

———. "Framing Strategic Deterrence: Old Certainties, New Ambiguities." *RUSI Journal* 154, no. 4 (2009).

———, and Efriam Karsh. *The Gulf Conflict 1990–1991.* Princeton: Princeton University Press, 1993.

Frisch, Hillel. "Motivation or Capabilities? Israeli Counterterrorism against Palestinian Suicide Bombings and Violence." *The Journal of Strategic Studies* 29, no. 5 (2006).

Ganor, Boaz. *The Counter-Terrorism Puzzle: A Guide for Decision Makers.* New York: Transaction Publishers, 2011.

Gartzke, Erik, and Jon Lindsay. "Weaving Tangled Webs: Offense, Defense, and Deception in Cyberspace." *Journal of Security Studies* 24 no. 2 (2015).

Gearson, John. "Deterring Conventional Terrorism: From Punishment to Denial and Resilience." *Contemporary Security Policy* 33, no. 1 (2012).

Geer, Daniel, Rebecca Bace, Peter Gutmann, Perry Metzger, Charles Pfleeger, John Quarterman, and Bruce Schneier. *CyberInsecurity: The Cost of Monopoly.* Washington DC: Computer and Communications Industry Association, 2003.

Geipel, Gary. "Urban Terrorists in Continental Europe after 1970: Implications for Deterrence and Defeat of Violent Nonstate Actors." *Comparative Strategy* 26, no. 5 (2007).

George, Alexander. "Coercive Diplomacy." In *The Use of Force,* edited by Robert Art and Kenneth Waltz. Lanham: Rowman and Littlefield, 2003.

———, and Richard Smoke. *Deterrence in American Foreign Policy.* New York: Columbia University Press, 1974.

Gerson, Michael. "Conventional Deterrence in the Second Nuclear Age." *Parameters,* 39, no. 3 (2009).

Giap, Vo Nguyen. "The Big Victory, the Great Task." In *Visions of Victory: Selected Vietnamese Communist Military Writings 1964 – 1968,* by Patrick McGarvey. Stanford: Hoover Institution on War, Revolution and Peace, 1969.

Gibbs, Jack. *Crime, Punishment, and Deterrence.* New York: Elsevier, 1975.

Glaser, Charles. "Deterrence of Cyber Attacks and U.S. National Security." Working Paper, George Washington University: Cyber Security Policy and Research Institute (2011).

———. "Why do Strategists Disagree about the Requirements of Strategic Nuclear Deterrence." In *Nuclear Arguments: Understanding the Strategic Nuclear Arms and Arms Control Debate*, edited by Lynn Eden and Steven Miller. Ithaca: Cornell University Press, 1989.

Goldstein, Avery. *Deterrence and Security.* Stanford: Stanford University Press, 2007.

Gray, Colin. *The Strategy Bridge: Theory for Practice.* Oxford: Oxford University Press, 2010.

Greenberg, David. "Studying New York City's Crime Decline: Methodological Issues." *Justice Quarterly* 31, no. 1 (2014).

Grogger, Jeffrey. "The Effects of Civil Gang Injunctions on Reported Violent Crime: Evidence from Las Angeles County." *Journal of Law and Economics* 45, no. 1 (2002).

Guerette, Rob, and Kate Bowers. "Assessing the Extent of Crime Displacement and Diffusion of Benefits." *Criminology* 47, no. 4 (2009).

Gutsell, Jennifer, and Michael Inzlicht, "Empathy Constrained: Prejudice Predicts Reduced Mental Simulation of Actions During Observation of Outgroups." *Journal of Experimental Social Psychology* 46, no. 5 (2010).

Hammes, Thomas. "War Evolves into Fourth Generation." In *Global Insurgency and the Future of Armed Conflict*, edited by Terry Terriff, Aaron Karp and Regina Karp. London: Routledge, 2008.

Handel, Michael. "The Evolution of Israeli Strategy." In *The Makers of Strategy*, edited by Williamson Murray, McGregor Knox, and Alvin Bernstein. Cambridge: Cambridge University Press, 1995.

Harcourt, Bernard. *Illusion of Order: The False Promise of Broken Windows Policing.* Cambridge, MA: Harvard University Press, 2009.

———. "Reflecting on The Subject: A Critique of The Social Influence Conception of Deterrence, The Broken Windows Theory, And Order-Maintenance Policing New York Style." *Michigan Law Review* 97 (1998).

———, and Jen Ludwig. "Broken Windows: New Evidence From New York City and A Five-City Experiment." *University of Chicago Law Review* 73, no. 1 (2006).

Harel, Amos. "Kah terae milhama be 2025." *Haaretz*, October 11, 2013.

Harknett, Richard. "The Logic of Conventional Deterrence and the End of the Cold War." *Security Studies* 4, no. 1 (1994).

———. "State Preferences, Systemic Constraints, and the Absolute Weapon." In *The Absolute Weapon Revisited*, edited by TV Paul, Richard Harknett, and James Wirtz. Ann Arbor: University of Michigan Press, 1998.

Harvey, Frank. "Rigor Mortis or Rigor, More Test: Necessity, Sufficiency, and Deterrence Logic." *International Studies Quarterly* 42, no. 4 (1998).

———, and Alex Wilner. "Counter-Coercion, the Power of Failure, and the Practical Limits of Deterring Terrorism." In *Deterring Terrorism: Theory and Practice*, edited by Andreas Wenger and Alex Wilner. Stanford: Stanford University Press, 2012.

Hendel, Yoez. "Shuvo shel maarach hamiluim." *Strategic Update* 10, no. 4 (February 2008).

Hennessey, Susan. "Deterring Cyberattacks: How to Reduce Vulnerability." *Foreign Affairs* 6, no. 6 (2017).

Henriksen, Dag "Deterrence by Default?" *Journal of Strategic Studies* 35, no. 1 (2012): 95–120.

Herley, Cormac. "When Does Targeting Make Sense for an Attacker?" *IEEE Security & Privacy* 11, no. 2 (2013).

Hipp, John. "Resident Perceptions of Crime and Disorder: How Much Is "Bias," and How Much Is Social Environment Differences?" *Criminology* 48 (2010).

Hirschi, Travis. *Causes of Delinquency*. University of California Press, 1969.

Hoffman, Wyatt, and Tristan Volpe. "Internet of Nuclear Things; Managing the Proliferation Risks of 3-D Printing Technology." *Bulletin of the Atomic Scientists* 74, no. 2 (2018).

Horowitz, Michael. "Nonstate Actors and the Diffusion of Innovations: The Case of Suicide Terrorism." *International Organization* 64, no. 1 (2010).

Hough, Mike, et al. "Procedural Justice, Trust, And Institutional Legitimacy." *Policing* 4 (2010).

Howell, Christian Jordan-Michael. "The Restrictive Deterrent Effect of Warning Banners in a Comprised Computer System." Graduate Thesis: University of South Florida, 2016.

Huntington, Samuel. "Conventional Deterrence and Conventional Retaliation in Europe." *International Security* 8, no. 3 (1983/4).

Huq, Aziz, et al. "Mechanisms for Eliciting Cooperation in Counter-Terrorism Policing: Evidence from the United Kingdom." *Journal of Empirical Legal Studies* 8, no. 4 (2011).

———, et al. "Why Does the Public Cooperate with Law Enforcement? The Influence of The Purposes and Targets Of Policing." *Psychology Public Policy and Law* 17 (2011).

Huth, Paul, and Bruce Russett. "General Deterrence between Enduring Rivals: Testing Three Competing Models." *American Political Science Review* 87, no. 1 (1993).

———. "Testing Deterrence Theory: Rigor Makes a Difference." *World Politics* 42, no. 4 (1990).

Innes, Martin. "Policing Uncertainty: Countering Terror Through Community Intelligence and Democratic Policing." *Annals of the American Academy of Political and Social Science* 605, no. 1 (2006).

Jackson, Brian, Peter Chalk, Kim Cragin, Bruce Newsome, John Parachini, William Rosenau, Erin Simpson, Melanie Sisson, and Donald Temple. *Breaching the Fortress Wall: Understanding Terrorist Efforts to Overcome Defensive Technologies.* Santa Monica, CA: RAND, 2007.

Jackson, Jonathan, et al. *Just Authority? Trust in the Police in England and Wales.* New York: Routledge, 2012.

Jackson, Jonathan. "A Psychological Perspective on Vulnerability in the Fear of Crime." *Psychology, Crime and Law* 15, no. 4 (2009).

Jacques, Scott, and Andrea Allen. "Bentham's Sanction Typology and Restrictive Deterrence." *Journal of Drug Issues* 44, no. 2 (2013).

Japanese Ministry of Defense. *Defense of Japan 2017.* October 2017.

———. *National Defense Program Guidelines for FY 2014 and beyond.* 2013.

———. *Order for Operation of the Self-Defense Forces Concerning Measures to Destroy Ballistic Missiles or Other Objects.* March 2009.

Jarman, Neil. *BIP Interface Mapping Project.* Belfast, Ireland: Institute for Conflict Research, 2005.

Johnson, David, Karl Mueller, and William Taft. *Conventional Coercion across the Spectrum of Conventional Operations* (No. 1494). Santa Monica: RAND, 2002.

Kahan, Dan. "Between Economics and Sociology: The New Path of Deterrence." *Michigan Law Review* 95 (1997).

Kaneda, Hideaki, Hiroshi Tajima, Kazumasa Kobayashi, and Hirofumi Tosaki. *Japan's Missile Defense: Diplomatic and Security Policies in a Changing Strategic Environment.* Tokyo: Japan Institute of International Affairs, 2007.

Kang, David, and Ji-Young Lee, "Japan-Korea Relations: Pyongyang's Belligerence Dominates." *Comparative Connections* 11:2 (2009).

Kaplan, Edward, Alex Mintz, Shaul Mishal and Claudio Samban. "What Happened to Suicide Bombings in Israel? Insights from a Terror Stock Model," *Studies in Conflict & Terrorism* 28, no. 3 (2005).

Kapur, S. Paul. "Deterring Nuclear Terrorists." In *Complex Deterrence*, edited by TV Paul, Pat Morgan, and Jim Wirtz. Chicago: University of Chicago Press, 2009.

Karmen, Andrew. *New York Murder Mystery: The True Story Behind the Crime Crash of the 1990s.* New York: New York University Press, 2000.

Kelling, George, and James Wilson. "Broken Windows: The Police and Neighborhood Safety." *The Atlantic*, March 1, 1982.

———, and William Sousa. "Do Police Matter? An Analysis of The Impact of New York City's Police Reforms." Manhattan Institute (2001).

———, and Mark Moore. "The Evolving Strategy of Policing." *Perspectives on Policing*. Washington, DC: National Institute of Justice, 1988.

Kennedy, David. *Deterrence and Crime Prevention*. New York: Routledge, 2009.

———. *Don't Shoot: One Man, A Street Fellowship, and the End of Violence in Inner-City America*. New York: Bloomsbury, 2011.

———. "Drugs, Race and Common Ground: Reflections on The High Point Intervention." *National Institute of Justice Journal* 262 (2009).

———. "Practice Brief: Norms, Narratives, and Community Engagement for Crime Prevention." John Jay College of Criminal Justice, (2010). Available at https://nnscommunities.org/wp-content/uploads/2017/1 0/Haas__practice_brief_finalwinter2010.pdf

———. Presentation at the 5th Annual Restorative Justice Initiative Conference (November 11, 2008).

Khalid, Kiran. "Some Experts See Fatwa as Significant Blow to Terrorist Recruiting." CNN, March 3, 2010.

Khan, Herma. *On Escalation: Metaphors and Scenarios* (3rd ed.). London: Routledge, 2017.

Kirk, David, and Mauri Matsuda. "Legal Cynicism, Collective Efficacy, and the Ecology of Arrest." *Criminology* 49 (2011).

Kirk, Davis, and Andrew Papachristos. "Cultural Mechanisms and The Persistence of Neighborhood Violence" *American Journal of Sociology* 116 (2011).

Kissinger, Henry. *On China*. New York: Penguin, 2011.

Kliot, Nurit, and Igal Charney. "The Geography of Suicide Terrorism in Israel." *GeoJournal* 66, no. 4 (2006).

Knopf, Jeff. "Terrorism and the Fourth Wave in Deterrence Research." In *Deterring Terrorism: Theory and Practice*, edited by Andreas Wenger and Alex Wilner. Stanford: Stanford University Press, 2012.

———. "The Fourth Wave in Deterrence Research." *Contemporary Security Policy*, 31, no. 1 (2010).

Kroenig, Matthew, and Barry Pavel. "How to Deter Terrorism." *Washington Quarterly*, 35, no. 2 (2012).

La Vigne, Nancy, et al. "Evaluating the Use of Public Surveillance Cameras for Crime Control and Prevention: A Summary." Washington, DC: Urban Institute, 2011.

Lafree, Gary, and Gary Ackerman. "The Empirical Study of Terrorism: Social and Legal Research." *Annual Review of Law and Social Science* 5 (2009).

———, and James Hendrickson. "Build a Criminal Justice Policy for Terrorism." *Criminology and Public Policy* 6, no. 4 (2007).

———, and Laura Dugan. "Introducing the Global Terrorism Database." *Terrorism and Political Violence* 19, no. 2 (2007).

———. "Expanding Criminology's Domain: The American Society of Criminology 2006 Presidential Address." *Criminology* 45 (2007).

Lake, David. "Rational Extremism: Understanding Terrorism in the Twenty-First Century." *Dialogue IO* 1, no. 1 (2002).

Lanoszka, Alexander, and Michael Hunzeker. "Confronting the Anti–Access/Area Denial and Precision Strike Challenge in the Baltic Region." *The RUSI Journal* 161, no. 5 (2016).

Lanoszka, Alexander and Michael Hunzeker. *Conventional Deterrence and Landpower in Northeastern Europe*. Carlisle, PA: US Army War College Press, 2019.

Larkin, Sean. "The Age of Transparency: International Relations Without Secrets." *Foreign Affairs* 95, no. 3 (2016).

Lebow, Richard Ned. "Thucydides and Deterrence." *Security Studies* 16 (April–June 2007).

Levi, Ron. "Making Counter-Law: On Having No Apparent Purpose in Chicago." *British Journal of Criminology* 49 (2009).

Libicki, Martin. *Conquest in Cyberspace*. Cambridge: Cambridge University Press, 2007.

———. *Cyberdeterrence and Cyberwar.* Washington, DC: RAND, 2009.

———. *Cyberspace in Peace and War.* Annapolis, MD: Naval Institute Press, 2016.

Lindsay, Jon, and Erik Gartzke, eds. *Cross-Domain Deterrence.* Oxford: Oxford University Press, 2019.

Lieberman, Elli. "Deterrence Theory: Successes or Failure in Arab-Israeli Wars?" *NDU McNair Papers*, no. 45 (October 1995).

———. *Reconceptualizing Deterrence: Nudging Toward Rationality in Middle Eastern Rivalries.* New York: Routledge, 2013.

Livingston, Debra. "Police Discretion and The Quality of Life in Public Places: Courts, Communities, and the New Policing." *Columbia Law Review* 97 (1997).

Long, Jerry Mark, and Alex Wilner, "Deterring an 'Army Whose Men Love Death': Delegitimizing al-Qaida." *International Security* 39, no. 1 (2014).

Lowther, Adam, ed. *Deterrence.* New York: Palgrave, 2012.

Lum, Cynthia, and Leslie Kennedy, eds. *Evidence-based Counterterrorism Policy.* London: Spring, 2012.

Lupovici, Amir. "The Emerging Fourth Wave of Deterrence Theory: Toward a New Research Agenda." *International Studies Quarterly* 54 (2010).

———. "Cyber Warfare and Deterrence." *INSS Military and Strategic Affairs*, 3:3 (2011): 32–33.

Lutes, Charles, and M. Elaine Bunn. "Dissuasion and the War on Terror: What is Meant by Dissuasion, and How Might It Apply to the War on Terror?" In *Five Dimensions of Homeland and International Security*, edited by Esther Brimmer. Washington, DC: Center for Transatlantic Relations, 2008.

Mandel, Robert. *Optimizing Cyberdeterrence: A Comprehsneive Strategy for Preventing Foreign Cyberattacks.* Washington, DC: Georgetown University Press, 2017.

Manyin, Mark. *North Korea-Japan Relations: The Normalization Talks and the Compensation/Reparations Issue.* Washington, DC: Congressional Research Service, 2002.

Matray, James. "Dean Acheson's Press Club Speech Reexamined," *Journal of Conflict Studies* 22, no. 1 (2002).

Mauroni, Albert. *Chemical-Biological Defense: US Military Policies and Decision in the Gulf War.* Westport, CT: Praeger, 1998.

Mazanec, Brian, and Bradley Thayer. *Deterring Cyber Warfare.* London: Palgrave, 2015.

McCarthy, Niall. "The Worrying Escalation of North Korea's Missile Tests," *The Statistics Portal,* July 2017, https://www.statista.com/chart/9172/the-worrying-escalation-of-north-koreas-missile-tests/.

McClure, Stuart, Joel Scambray, and George Kurtz. *Hacking Exposed: Network Security Secrets & Solutions* (7th ed.). New York: McGraw-Hill, 2012.

McKittrick, David, and David McVea. *Making Sense of the Troubles: The Story of the Conflict in Northern Ireland.* New York: New Amsterdam Books, 2002.

Meares, Tracey and Dan Kahan. "Law and (Norms Of) Order in The Inner City." *Law and Society Review* 32, no. 4 (1998).

Mearsheimer, John. "Prospects for Conventional Deterrence in Europe." *Bulletin of the Atomic Scientists* 41, no. 7 (1985).

———. "Why the Soviets Can't Win Quickly in Central Europe" *International Security* 7, no. 1 (1982).

———. *Conventional Deterrence.* Ithaca: Cornell University Press, 1983.

Merari, Ariel. Testimony before the Special Oversight Panel on Terrorism. U.S. House of Representatives (July 13, 2000).

Mesev, Victor, Joni Downs, Aaron Binns, Richard Courtney, and Peter Shirlow. "Measuring and Mapping Conflict-Related Deaths and Segregation: Lessons from the Belfast 'troubles.'" in *Geospatial Technologies and Homeland Security*, edited by Daniel Z. Sui. Dordrecht: Springer, 2008.

Messner, Steven, et al. "Policing, Drugs, and The Homicide Decline in New York City in the 1990s." *Criminology* 45 (2007).

Metz, Cade. "China's Blitz to Dominate A.I." *New York Times.* February 13, 2018.

Missile Defense Advocacy Alliance. "Making the World a Safer Place: Japan." June 2018.

Mizokami, Kyle. "Everything You Need to Know: Japan's Missile Defenses." *National Interest,* September 2, 2017.

Maoz, Zeev. *Defending the Holy Land.* Ann Arbor: University of Michigan Press, 2009.

Moghadam, Assaf. "Motives for Martyrdom: Al-Qaida, Salafi Jihad, and the Spread of Suicide Attacks." *International Security* 33 (2009).

Morenoff, Jeffrey, et al. "Neighborhood Inequality, Collective Efficacy, and The Spatial Dynamics of Urban Violence." *Criminology* 39 (2001).

Morgan, Patrick. "The State of Deterrence in International Politics Today." *Contemporary Security Policy* 33, no. 1 (2012).

———. *Deterrence: A Conceptual Analysis.* Beverly Hills, CA: Sage, 1983.

———. *Deterrence Now.* Cambridge: Cambridge University Press, 2003.

Morral Andrew, and Brian Jackson. "Understanding the Role of Deterrence in Counterterrorism Security" *RAND Occasional Paper* (2009).

Nagin, Daniel. "Deterrence in the Twenty-First Century." *Crime and Justice* 42, no. 1 (2013).

National Consortium for the Study of Terrorism and Responses to Terrorism. Global Terror Database (2020).

National War College. *Combating Terrorism in a Globalized World: Report by the NWC Student Task Force on Combating Terrorism.* May 2002.

North Atlantic Treaty Organization (NATO), Standardization Agency (NSA). *NATO Glossary of Terms and Definitions* (2008).

———. Press Conference with Secretary General Jens Stoltenberg (June 14, 2016).

Nuclear Threat Initiative. *Sohae Satellite Launching Station.* December 2012.

Nye, Joseph. "Deterrence and Dissuasion in Cyberspace." *International Security* 41, no. 3 (2017).

O'Donnell, Frank, and Yogesh Joshi. "India's Missile Defense: Is the Game Worth the Candle?" *The Diplomat,* August 2, 2013.

Obama, Barack. "Remarks of President Barack Obama." Prague: Embassy of the United States, April 2009.

Oren, Amir. "Et hadegel al hagiva itkeu hametosim." *Haaretz,* January 31, 2013.

Overy, R. J. "Air Power and the Origins of Deterrence Theory before 1939." *Journal of Strategic Studies* 15, no.1 (1992): 73–101

Palita, Zohar. "Israel's Security Fence: Effective in Reducing Suicide Attacks from the Northern West Bank." *Policywatch* No. 464 (2004).

Park, Cheol Hee. "Japanese Strategic Thinking toward Korea." In *Japanese Strategic Thought toward Asia,* edited by Gilbert Rozman, Kazuhiko Togo, and Joseph Ferguson. New York: Palgrave Macmillan, 2007.

Paul, TV, Patrick Morgan, and Jim Wirtz, eds. *Complex Deterrence.* Chicago: University of Chicago Press, 2009.

Payne, Keith. "Deterrence and Coercion of Non-State Actors: Analysis of Case Studies." Fairfax, VA: National Institute for Public Policy, 2008.

———. *Deterrence in the 2nd Nuclear Age.* Lexington: University of Kentucky, 1996.

———. *Strategy, Evolution, and War: From Apes to Artificial Intelligence.* Washington, DC: Georgetown University Press, 2018.

———. *The Great American Gamble: Deterrence Theory and Practice from the Cold War to the Twenty–First Century.* Fairfax, Virginia: National Institute Press, 2008.

Pearlman, Wendy, and Boaz Atzili. *Triadic Coercion: Israel's Targeting of States that Host Nonstate Actors.* New York: Columbia University Press, 2019.

Pedatzur, Reuven. "How Missile Defense Undermines Deterrence: The Israeli Case." INSS, January 2014.

Perrow, Charles. *Normal Accidents.* 2nd ed. New Haven, Conn: Yale University Press, 1999.

Perrow, Charles. *The Next Catastrophe: Reducing our Vulnerabilities to Natural, Industrial, and Terrorist Disasters.* Princeton: Princeton University, 2007.

Perry, Charles, Jacquelyn Davis, James Schoff, and Toshi Yoshihara. *Alliance Diversification & the Future of the U.S.–Korean Security Relationship.* Dulles: Brassey's, Inc., 2004.

Petrelli, Niccolo. "Deterring Insurgents: Culture, Adaptation and the Evolution of Israeli Counterinsurgency." *Journal of Strategic Studies* 35, no. 1 (2013).

Pinkston, Daniel. *The North Korean Ballistic Missile Program.* Carlisle, PA: Strategic Studies Institute, 2008.

Posen, Barry. "U.S. security policy in a nuclear-armed world or: What if Iraq had had nuclear weapons?" *Security Studies* 6:3 (1997).

———. "US Security Policy in a Nuclear-Armed World, or What if Iraq had had Nuclear Weapons?" In *The Coming Crisis: Nuclear Proliferation, US Interests, and World Order*, edited by Victor Utgoff. Cambridge: Belfer Center, 2000.

Post, Jerrold. "Terrorist Psycho-Logic: Terrorist Behavior as a Product of Psychological Forces." In *Origins of Terrorism: Psychologies, Ideologies, Theologies, States of Mind*, edited by Walter Reich. Washington, DC: Woodrow Wilson Center Press, 1998.

Presidential Decision Directive 39. *U.S. Policy on Counterterrorism.* Washington DC: United States White House Office, 1995.

Prudente, Tim. "Seeing stars, again: Naval Academy reinstates celestial navigation." *Capital Gazette*, October 12, 2015.

Quester, George. *Deterrence Before Hiroshima: The Airpower Background of Modern Strategy.* Piscataway, NJ: Transaction Books, 1986.

Rapoport, Amir. "Hatkifa bein hamilhamot." *Maariv*, February 1, 2013.

Ratcliffe, Jerry, and Clarissa Breen. "Crime Diffusion and Displacement: Measuring the Side Effects of Police Operations." *The Professional Geographer* 63, no. 2 (2011).

Ratcliffe, Jerry, et al. "The Crime Reduction Effects of Public CCTV Cameras." *Justice Quarterly* 26, no. 4 (2009).

Raudenbush, Stephen, and Anthony Byrk. *Hierarchical Linear Models: Applications and Data Analysis Methods.* London: Sage, 2001.

Ravenscroft, Emily. "The Meaning of the Peacelines of Belfast." *Peace Review* 21, no. 2 (2009).

Reuters. "As North Korea missile threat grows, Japan lawmakers argue for first strike options." March 8, 2017.

Rhodes, Edward. *Power and MADness: The Logic of Nuclear Coercion.* New York: Columbia University Press, 1989.

Thomas. "Deterrence Beyond the State: The Israeli Experience." *Contemporary Security Policy* 33, no. 2 (2012).

Roach, Geoff. "For the Sake of Civility, Clean Out Hindley Sewer." *The Advertiser*, March 14, 2009.

Roberts, Brad. "Extended Deterrence and Strategic Stability in Northeast Asia." *Visiting Scholars Paper Series* 1, Tokyo: The National Institute for Defense Studies, 2013.

———. *The Case for U.S. Nuclear Weapons in the 21st Century.* Stanford: Stanford University Press, 2016.

Rosen, Stephen. *Winning the Next War.* Ithaca: Cornell University Press, 1991.

Rosenbaum, Dennis, ed. *The Challenge of Community Policing: Testing the Promises.* New York: Sage, 1994.

Rosenberg, David Alan. "U.S. Nuclear War Planning, 1945–1960." In *Strategic Nuclear Targeting*, edited by Desmond Ball and Jeffrey Richelson. Ithaca: Cornell University Press, 1986.

Rosendorff, B. Peter, and Todd Sandler. "Too Much of a Good Thing? The Proactive Response Dilemma." *Journal of Conflict Resolution* 48, no. 5 (2004).

Rosenfeld, Richard, et al. "The Impact of Order-Maintenance Policing On New York City Homicide and Robbery Rates: 1988–2001." *Criminology* 45 (2007).

Rubin, Uzi. "Israel's Missile Defense: An Asset or a Drawback in a Nonconventional Scenario." INSS, January 2014

Sampson, Robert, and Dawn Bartusch. "Legal Cynicism And (Subcultural?) Tolerance of Deviance: The Neighborhood Context of Racial Differences." *Law and Society Review* 32, no. 4 (1998).

———, and Stephen Raudenbush. "Systematic Social Observation of Public Spaces: A New Look at Disorder in Urban Neighborhoods." *American Journal of Sociology* 105, no. 3 (1999).

———, and Steve Raudenbush, "Disorder in Urban Neighborhoods: Does It Lead to Crime?" National Institute of Justice, Research Brief 5 (2001).

———, et al., "Beyond Social Capital: Spatial Dynamics of Collective Efficacy for Children." *American Sociological Review* 64, no. 5 (1999).

———, et al., "Neighborhoods and Violent Crime: A Multilevel Study of Collective Efficacy," *Science* 277, no. 5328 (1997).

———. "Moving and The Neighborhood Glass Ceiling." *Science* 337 (2012).

———. "The Place of Context: A Theory and Strategy for Criminology's Hard Problems." *Criminology* 51, no. 1 (2013).

———. *Great American City: Chicago and the Enduring Neighborhood Effect.* Chicago: University of Chicago Press, 2012.

Sandler, Todd, and Kevin Siqueira. "Global Terrorism: Deterrence versus Pre-Emption." *Canadian Journal of Economics/Revue Canadienne D'économique* 39:4 (2006).

Sandler, Todd. "Collective Versus Unilateral Response to Terrorism." *Public Choice* 24 (2005).

Sartori, Ann. *Deterrence by Diplomacy.* Princeton: Princeton University Press, 2005.

Schachter, Jonathan. "Unusually Quiet: Is Israel Deterring Terrorism?" *Strategic Assessment* 13, no. 2 (2010).

Schaub, Gary, Jr. "When is Deterrence Necessary: Gauging Adversary Intent." *Strategic Studies Quarterly* 3, no. 4 (2009).

Schelling, Thomas. *Arms and Influence.* New Haven: Yale University Press, 1966.

———. *Arms and Influence.* 2nd ed. New Haven: Yale University Press, 2008.

———. *The Strategy of Conflict.* Cambridge: Harvard University Press, 1980.

Schiller, Markus. *Characterizing the North Korean Nuclear Missile Threat.* Santa Monica: RAND Corporation, 2012.

Schneider, Barry. "Deterrence and Saddam Hussein." The Counterproliferation Papers, Future Warfare Series No. 47. Maxwell AFB: USAF Counterproliferation Center, 2009.

Schneier, Bruce. "Refuse to be Terrorized." *Schneier on Security,* August 24, 2006, https://www.schneier.com/essays/archives/2006/08/refuse_to_be_terrori.html.

Shalal, Andrea. "U.S. firm CrowdStrike claims success in deterring Chinese hackers." *Web Culture,* April 14, 2015.

Shapiro, Ian. *Containment.* Princeton: Princeton University Press, 2007.

Shimshoni, Jonathan. *Israel and Conventional Deterrence: Border Warfare from 1953 to 1970.* Cornell: Cornell University Press, 1988.

Shirlow, Peter, and Joni Downs. "The Geography of Conflict and Death in Belfast, Northern Ireland." *Annals of the Association of American Geographers* 99, no. 5 (2009).

Singer, Peter, and Allan Friedman. *Cybersecurity and Cyberwar.* Oxford: Oxford University Press, 2014.

Siqueira, Kevin. "Political and Militant Wings within Dissident Movements and Organizations." *The Journal of Conflict Resolution* 49, no. 2 (2005).

Skogan, Wesley, and Kathleen Frydl, eds. *Fairness and Effectiveness in Policing: The Evidence.* Washington, DC: National Academies Press, 2004.

———. "Broken Windows: Why—And How—We Should Take Them Seriously." *Criminology and Public Policy* 7 (2008).

———. *Disorder and Decline: Crime and the Spiral of Decay in American Neighborhoods.* University of California Press, 1990.

Slater, Jerome. "Just War Model Philosophy and the 2008–2009 Israeli Campaign in Gaza." *International Security* 37, no. 2 (2012): 44–80.

Slaughter, Anne-Marie. *The Chessboard and the Web: Strategies of Connection in a Networked World.* New Haven: Yale University Press, 2017.

Slayton, Rebecca. "What Is the Cyber Offense-Defense Balance? Conceptions, Causes, and Assessment." *International Security* 41, no. 3 (2017).

Smith, Derek. *Deterring America: Rogue States and the Proliferation of Weapons of Mass Destruction.* New York: Cambridge University Press, 2006.

Smith, James, and Brent Talbot. "Terrorism and Deterrence by Denial." In *Terrorism and Homeland Security*, edited by Paul Viotti, Michael Opheim, and Nicholas Bowen. New York: CRC Press, 2008.

Smith, James. "Strategic Analysis, WMD Terrorism, and Deterrence by Denial." In *Deterring Terrorism: Theory and Practice*, edited by Andreas Wenger and Alex Wilner. Stanford: Stanford University Press, 2012.

Snyder, Glenn. "Deterrence and Defense." Reprinted in *The Use of Force: International Politics and Foreign Policy*, edited by Robert Art and Kenneth Waltz. New York: University Press of America, 1983.

———. *Deterrence and Defense: Toward a Theory of National Security.* Princeton: Princeton University Press, 1961.

———. "Deterrence and Power." *Journal of Conflict Resolution* 4, no. 2 (1960).

Snyder, Jack. *Soviet Strategic Culture: Implications for Limited Nuclear Operations.* Santa Monica: RAND, 1977.

Spaniel, William. "Rational Overreaction to Terrorism." *Journal of Conflict Resolution* 63, no. 3 (2019).

Spilerman, Seymour and Guy Stecklov. "Societal Responses to Terrorist Attacks," *Annual Review of Sociology* 35 (2009).

Sri Bhashyam, Sumitra, and Gilberto Montibeller. "In the Opponent's Shoes: Increasing the Behavioral Validity of Attackers' Judgments in Counterterrorism Models." *Risk Analysis* 36, no. 4 (2016).

Stafford, Mark, and Mark Warr. "A Reconceptualization of General and Specific Deterrence." *Journal of Research in Crime and Delinquency* 30, no. 2 (1993).

Stein, Janice Gross. "Deterrence and Compellence in the Gulf, 1990-91." *International Security* 17, no. 2 (1992).

———. "Deterrence and Learning in an Enduring Rivalry: Egypt–Israel, 1948–73." *Security Studies* 6, no.1 (1996): 104–152.

———. "Deterring Terrorism, Not Terrorists." In *Deterring Terrorism: Theory and Practice*, edited by Andreas Wenger and Alex Wilner. Stanford: Stanford University Press, 2012.

———, and Ron Levi. "The Social Psychology of Denial: Deterring Terrorism," *New York University Journal of International Law and Politics* 47 (2015).

———, and Ron Levi. "Testing Deterrence by Denial: Experimental Results from Criminology," *Studies in Conflict and Terrorism* (forthcoming).

Stenzler-Koblentz, Liram. "Iron Dome's Impact on the Military and Political Agenda: Moral Justifications for Israeli to Launch a Military Operation against and Terrorist and Guerilla Organizations." *Military and Strategic Affairs* 6, no. 1 (2014): 79–97.

Stevens, Tim. "A Cyberwar of Ideas? Deterrence and Norms in Cyberspace." *Contemporary Security Policy* 33, no. 1 (2012).

Stone, John. "Conventional Deterrence and the Challenge of Credibility." *Contemporary Security Policy* 33, no. 1 (2012).

Straits Times. "North Korea's Hwasong-12 is the 'most serious missile to watch,'" September 17, 2017.

Szyliowicz, Joseph. "Aviation Security: Promise or Reality?" *Studies in Conflict and Terrorism* 27, no. 1.

Takahashi, Sugio. "Ballistic Missile Defense in Japan: Deterrence and Military Transformation." *Asie.Visions* 59/*Proliferation Papers* 44 (2012).

Telegraph. "Underwear Bomber Plot Failed Because He 'Wore Same Pants for Two Weeks.'" July 25, 2014.

Terrill, W. Andrew. *Escalation and Intrawar Deterrence during Limited Wars in the Middle East.* Carlisle, PA: US Army War College Strategic Studies Institute, 2009.

Tertrais, Bruno. "Drawing Red Lines Right." *The Washington Quarterly* 37, no. 3 (2014).

Tetlock, Philip, and Ariel Levi. "Attribution Bias: On the Inconclusiveness of the Cognition-Motivation Debate." *Journal of Experimental Social Psychology* 18, no. 1 (1982).

Thomas, Troy, Stephen Kiser, and William Casebeer. *Warlord Rising: Confronting Violent Non-state Actors.* UK: Lexington, 2005.

Times of Israel. "David's Sling Success Caught on Film." November 27, 2012.

Tor, Uri. "'Cumulative Deterrence' as a New Paradigm for Cyber Deterrence." *Journal of Strategic Studies* 40, no. 1 (2017).

Toth, Robert. "American Support Grows for Use of Nuclear Arms." *LA Times,* February 3, 1991.

Trager, Robert, and Dessislava Zagorcheva. "Deterring Terrorism: It Can Be Done." *International Security* 30, no. 3 (2005/6).

Tucker, Jonathan. "Evidence Iraq Used Chemical Weapons During the 1991 Persian Gulf War." *The Nonproliferation Review* (Spring-Summer 1997).

———. *Strategies for Countering Terrorism: Lessons from The Israeli Experience* (2003), https://web.archive.org/web/20080723105821/http://www.homelandsecurity.org/journal/Articles/tucker-israel.html

Tyler, Tom. "Legitimacy and Criminal Justice: The Benefits of Self-Regulation." *Ohio State Journal of Criminal Law* 7 (2009).

———. "Restorative Justice and Procedural Justice: Dealing with Rule Breaking." *Journal of Social Issues* 62, no. 2 (2006).

———. "Toughness Vs. Fairness: Police Policies and Practices for Managing the Risk of Terrorism." In *Evidence-based Counterterrorism Policy*, edited by Cynthia M. Lum and Leslie W. Kennedy. New York: Springer, 2012.

———. "What Do They Expect? New Findings Confirm the Precepts of Procedural Fairness." *California Court Review* (Winter 2006).

———. *Why People Obey the Law.* Princeton: Princeton University Press, 2006.

———, et al., "Legitimacy and Deterrence Effects in Counterterrorism Policing: A Study of Muslim Americans." *Law and Society Review* 44, no. 2 (2010).

———, and Jeffrey Fagan. "Legitimacy and Cooperation: Why Do People Help the Police Fight Crime in Their Communities?" *Ohio State Journal of Criminal Law* 6 (2008).

———, and Yuen Huo. *Trust in The Law: Encouraging Public Cooperation with the Police and Courts.* New York: Russell Sage Foundation, 2002.

von Hlatky, Stéfanie and Andreas Wenger, eds. *The Future of Extended Deterrence: The United States, NATO, and Beyond.* Washington, DC: Georgetown University Press, 2015.

Wehling, Fred. "A Toxic Cloud of Mystery: Lessons from Iraq for Deterring CBRN Terrorism." In *Deterring Terrorism: Theory and Practice,* edited by Andreas Wenger and Alex Wilner. Stanford: Stanford University Press, 2012.

Weisburd, David, et al. "Displacement of Crime and Diffusion of Crime Control Benefits in Large-Scale Geographic Areas." Campbell Collaboration (2010).

———, et al. "Does Crime Just Move Around the Corner?" Criminology 44 (2006).

———, et al. "Policing, Terrorism, And Beyond." In *To Protect and To Serve: Policing In An Age Of Terrorism,* edited by David Weisburd, et al. New York: Springer, 2011.

———, et al. "Understanding and Controlling Hot Spots of Crime: The Importance of Formal and Informal Social Controls," *Prevention Science* 15, no. 1 (2014).

———, and John Eck. "What Can Police Do to Reduce Crime, Disorder, and Fear?" *Annals of the American Academy of Political and Social Science* 593 (2004).

Wenger, Andreas. "Conclusion: Reconciling Alliance Cohesion with Policy Coherence." In *The Future of Extended Deterrence: The United States, NATO, and Beyond*, edited by Stéfanie von Hlatky and Andreas Wenger. Washington, DC: Georgetown University Press, 2015.

———. *Living with Peril: Eisenhower, Kennedy, and Nuclear Weapons*. New York: Rowman & Littlefield, 1997.

———, and Alex Wilner, eds. *Deterring Terrorism: Theory and Practice*. Stanford: Stanford University Press, 2012.

———, and Alex Wilner. "Deterring Terrorism: Moving Forward." In *Deterring Terrorism: Theory and Practice*, edited by Andreas Wenger and Alex Wilner. Stanford: Stanford University Press, 2012.

Wilner, Alex. "Contemporary Deterrence Theory and Counterterrorism: A Bridge too Far?" *NYU Journal of International Law and Politics* 47, no. 2 (2015).

———. *Deterring Rational Fanatics*. Philadelphia: University of Pennsylvania Press: 2015.

———. "Deterring the Undeterrable: Coercion, Denial, and Delegitimization in Counterterrorism." *Journal of Strategic Studies* 34, no. 1 (2011).

———. "Fencing in Warfare: Threats, Punishment, and Intra-war Deterrence in Counterterrorism." *Security Studies* 22, no. 4 (2013).

———. "US Cyber Deterrence: Practice Guiding Theory." *Journal of Strategic Studies* 43, no. 2 (2020).

Wirtz, James. "Deterring the Weak: Problems and Prospects." *Proliferation Papers*, No. 43, 2012.

———. "Politics with Guns: A Response to T.X. Hammes's 'War evolves into the fourth generation.'" In *Global Insurgency and the Future of*

Armed Conflict, edited by Terry Terriff, Aaron Karp and Regina Karp. London: Routledge, 2008.

Wortley, Richard. "Guilt, Shame, And Situational Crime Prevention." In *The Politics and Practice of Situational Crime Prevention* (vol 5.), edited by Ross Homel. New York: Willow Tree Press, 1996.

Yadlin, Amos. "Time for Decisions." In *Strategic Survey for Israel*, edited by Shlomo Brom and Anat Kurtz. INSS, 2013.

Yamaguchi, Mari. "Japan to buy Aegis Ashore missile defense systems." *Associated Press*, December 19, 2017.

YouTube. "USENIX Enigma 2016 - NSA TAO Chief on Disrupting Nation State Hackers," https://www.youtube.com/watch?v=bDJb8WOJYdA. Accessed April 29, 2020.

Zagare, Frank. "Deterrence is Dead. Long Live Deterrence." *Conflict Management and Peace Science*, 23, no. 2 (2006).

Zimring, Franklin. *The City That Became Safe: New York's Lessons for Urban Crime and Its Control.* Oxford: Oxford University Press, 2011.

INDEX

ABOUT THE CONTRIBUTORS

Alex Wilner is Associate Professor of International Affairs at the Norman Paterson School of International Affairs, Carleton University, Ottawa, Canada.

Andreas Wenger is Professor of International and Swiss Security Policy at ETH Zurich and Director of the Center for Security Studies, Zurich, Switzerland.

Patrick M. Morgan is Professor of Political Science at the University of California, Irvine, USA.

Ron Levi is Associate Professor in the Munk School of Global Affairs & Public Policy and the Department of Sociology, University of Toronto, Canada.

Janice Gross Stein is the Belzberg Professor of Conflict Management in the Department of Political Science and Founding Director of the Munk School of Global Affairs and Public Policy, University of Toronto, Canada.

John Sawyer is the Director of Strategic Corporate Research Relationships at the National Consortium for the Study of Terrorism and Responses to Terrorism (START), a Department of Homeland Security Emeritus Center for Excellence led by the University of Maryland, USA.

James J. Wirtz is Dean of the School of International Graduate Studies, and Professor with the Department of National Security Affairs, at the Naval Postgraduate School, Monterey, USA.

Jonathan Trexel is (retired) Senior Policy Consultant at Science Applications International Corporation, USA.

Dmitry (Dima) Adamsky is Professor at the School of Government, Diplomacy and Strategy at the IDC Herzliya University, Israel.

Martin Libicki is Keyser Chair of Cybersecurity Studies at the US Naval Academy, Annapolis, USA.

CAMBRIA RAPID COMMUNICATIONS IN CONFLICT AND SECURITY (RCCS) SERIES

General Editor: Geoffrey R. H. Burn

The aim of the RCCS series is to provide policy makers, practitioners, analysts, and academics with in-depth analysis of fast-moving topics that require urgent yet informed debate. Since its launch in October 2015, the RCCS series has the following book publications:

- *A New Strategy for Complex Warfare: Combined Effects in East Asia* by Thomas A. Drohan
- *US National Security: New Threats, Old Realities* by Paul R. Viotti
- *Security Forces in African States: Cases and Assessment* edited by Paul Shemella and Nicholas Tomb
- *Trust and Distrust in Sino-American Relations: Challenge and Opportunity* by Steve Chan
- *The Gathering Pacific Storm: Emerging US-China Strategic Competition in Defense Technological and Industrial Development* edited by Tai Ming Cheung and Thomas G. Mahnken
- *Military Strategy for the 21st Century: People, Connectivity, and Competition* by Charles Cleveland, Benjamin Jensen, Susan Bryant, and Arnel David
- *Ensuring National Government Stability After US Counterinsurgency Operations: The Critical Measure of Success* by Dallas E. Shaw Jr.
- *Reassessing U.S. Nuclear Strategy* by David W. Kearn, Jr.
- *Deglobalization and International Security* by T. X. Hammes
- *American Foreign Policy and National Security* by Paul R. Viotti

- *Make America First Again: Grand Strategy Analysis and the Trump Administration* by Jacob Shively
- *Learning from Russia's Recent Wars: Why, Where, and When Russia Might Strike Next* by Neal G. Jesse
- *Restoring Thucydides: Testing Familiar Lessons and Deriving New Ones* by Andrew R. Novo and Jay M. Parker
- *Net Assessment and Military Strategy: Retrospective and Prospective Essays* edited by Thomas G. Mahnken, with an introduction by Andrew W. Marshall
- *Deterrence by Denial: Theory and Practice* edited by Alex S. Wilner and Andreas Wenger

For more information, visit www.cambriapress.com.

www.ingramcontent.com/pod-product-compliance
Lightning Source LLC
Chambersburg PA
CBHW031412270326
41929CB00010BA/1429